Genes and Environment in Personality Development

Sage Series on Individual Differences and Development

Robert Plomin, *Series Editor*

The purpose of the Sage Series on Individual Differences and Development is to provide a forum for a new wave of research that focuses on individual differences in behavioral development. A powerful theory of development must be able to explain individual differences, rather than just average developmental trends, if for no other reason than that large differences among individuals exist for all aspects of development. Variance—the very standard deviation—represents a major part of the phenomenon to be explained. There are three other reasons for studying individual differences in development: First, developmental issues of greatest relevance to society are issues of individual differences. Second, descriptions and explanations of normative aspects of development bear no necessary relationship to those of individual differences in development. Third, questions concerning the processes underlying individual differences in development are more easily answered than questions concerning the origins of normative aspects of development.

Editorial Board

Books in this series

Volume 1 HIGH SCHOOL UNDERACHIEVERS: What Do They Achieve as Adults?
Robert B. McCall, Cynthia Evahn, and Lynn Kratzer

Volume 2 GENES AND ENVIRONMENT IN PERSONALITY DEVELOPMENT
John C. Loehlin

Volume 3 THE NATURE OF NURTURE
Theodore D. Wachs

Genes and Environment in Personality Development

John C. Loehlin

Individual Differences and Development Series
VOLUME 2

SAGE Publications
International Educational Professional Publishers
Newbury Park London New Delhi

For information address:

SAGE Publications, Inc.
2455 Teller Road
Newbury Park, California 91320

SAGE Publications Ltd.
6 Bonhill Street
London EC2A 4PU
United Kingdom

SAGE Publications India Pvt. Ltd.
M-32 Market
Greater Kailash I
New Delhi 110 048 India

Printed in the United States of America

Library of Congress Cataloging-in-Publication Data

Loehlin, John C.
Genes and environment in personality development / John C. Loehlin.
p. cm. — (Sage series on individual differences and development ; vol. 2)
Includes bibliographical references and index.
ISBN 0-8039-4450-0 (cl). — ISBN 0-8039-4451-9 (pb)
1. Personality—Genetic aspects. 2. Personality development. 3. Nature and nurture. 4. Individual differences. 5. Behavior genetics. I. Title. II. Series.
BF698.9.B5L64 1992
155.7—dc20 92-3493
CIP

92 93 94 10 9 8 7 6 5 4 3 2 1

Sage Production Editor: Judith L. Hunter

Contents

Tables and Figures

Series Editor's Preface

A reasonable starting point in exploring the etiology of individual differences is quantitative genetics, which is called behavioral genetics when applied to the study of behavior. Behavioral genetics asks the extent to which genetic differences among individuals contribute to the differences we observe in their behavior. This is the issue of nature and nurture, one of the oldest questions in the behavioral sciences.

John C. Loehlin is a leader, if not *the* leader, in the field of human behavioral genetics. He is past president of the International Behavior Genetics Association, and this past summer he received the association's highest award, the Dobzhansky Award for an outstanding research career in behavioral genetics. His award citation noted that in his hands, the powerful tools of behavioral genetics have taken us far beyond the nature-nurture argument about the relative contributions of genetic and environmental influence. He has a special gift for taking the most difficult concepts of behavioral genetics and presenting them clearly, concisely, and with gusto.

John Loehlin is also the author of a 1976 book on the genetics of personality that reported several important discoveries that remain central topics of research on the genetics of personality. Loehlin's new book on the genetics of personality is not an update of his classic book;

it is a treasure of new approaches and new findings destined to become another landmark for the field. Loehlin wields with panache the powerful new tools of latent variable model fitting. Model fitting—also called causal, structural, biometrical, or path modeling—tests the fit between an explicit model and observed data. Scarcely known in 1976, model fitting is now de rigueur in reporting behavioral genetic research. Model fitting is important because it tests the adequacy of the model and its assumption, analyzes all information simultaneously, weights each piece of information according to its sample size, yields parameter estimates and standard errors for the best-fitting model, and compares alternative models.

Model fitting is never clearer than when John Loehlin writes about it. He is especially good at showing us how to get the most from model fitting's underappreciated power as a convenient and elegant means of summarizing and synthesizing large amounts of data. Most admirable is the way he deftly uses these tools to cut through the complexities of nature and nurture to make important substantive discoveries. For example, one of the many exciting conclusions drawn from his analyses is that genes contribute appreciably to developmental change, not just to continuity. For the newcomer to research on the genetics of personality, the book will help to "give the field away," so that these research strategies can be understood and used by researchers who are not behavioral geneticists. For the expert, the book offers many important findings that will set the agenda for research on the genetics of personality for the next decade.

ROBERT PLOMIN

Preface

This book is intended as an introduction to the behavior genetics of personality, with some special focus on issues of development and change. A person coming to this topic with information from, say, the general textbooks of 10 years ago, which were based on the specialized reviews of 13 years ago, which were in turn based on the studies of 15-plus years ago, may be surprised at how much evidence is now available on some points—and also at how little progress has been made on some others.

Strict limits on length have necessitated that this book be an introduction, rather than a comprehensive treatment. Thus I have had to skip many interesting topics; if your favorites have been slighted in the process, I apologize. I have tried to go into enough detail on what I do cover to give the reader some real sense of the articulation of conclusions with data, and the flexibility and power of modern model-fitting methods for accomplishing this.

My thanks go to Hill Goldsmith and Robert Plomin and the students in their respective seminars for critiquing an initial draft of the book. Lee Willerman also provided helpful comments. As usual, these individuals should be credited for the errors and obfuscations that got taken out (which you can't see) and not blamed for the ones that got left in (which you can). For permission to adapt previously published material, I thank the Institute for Research in Child Development, the American Psychological Association and Kathleen McCartney, the Academic Press and Robert Plomin and Gerald McClearn, and the Guilford Press, and for permission to use unpublished material, Sandra Scarr.

Finally, I would like to dedicate this book to two major pioneers in the behavior genetics of personality: Steven G. Vandenberg and Raymond B. Cattell. Whether they intended to or not, they helped get me started in this field.

J. C. L.

Introduction

The focus of this book will be on differences among individuals and how these differences come about. Not all differences: we will not be considering height, weight, blood pressure, or skin color; nor cognitive, athletic, or artistic skill. Rather, we will be concerned with the range of human expressive characteristics commonly designated by such terms as personality and temperament, and we will be examining their genetic and environmental causes.

Our emphasis will be on differences, that is, on how and why Amy is more sociable than Mary, Joe more active than Fred, Donna more emotionally stable than Louise. Some writers have speculated about how genes and environments in evolutionary history have led to basic human nature—those temperamental and personality features that Amy, Mary, Joe, Fred, Donna, and Louise (and all other normal humans) share. We will consider this matter in our final chapter; but for now, individual differences and their causes will be our theme.

Much of the empirical research on this topic in recent decades has gone under the label of "behavior genetics." A special subdivision, developmental behavioral genetics, is sometimes defined (e.g., Plomin, 1986). In a way, this terminology is redundant, since all research addressing the genetic and environmental origins of behavior is, in a broad sense, developmental. Nonetheless, the narrower term does have the virtue of focusing attention on certain issues within the wider field, such as the possibility of different contributions of genes and environment to personality variation at different points in the life span, or the roles that genes and environment play in causing personality changes over time—issues with which we will be concerned later in this book.

Of course, what will be said in these pages leans heavily on what has gone before. Many topics which will necessarily be covered briefly here may be pursued at more length in the specific sources cited throughout the text and included in the reference list at the end of the book, as well as in general behavior genetics textbooks such as those by Fuller and Thompson (1978), Hay (1985), and Plomin, DeFries and McClearn (1990), or in more specialized books such as *Development, Genetics, and Psychology* by Plomin (1986), *Genes, Culture and Personality* by Eaves, Eysenck, and Martin (1989), and Nichols and my *Heredity, Environment, and Personality* (Loehlin & Nichols, 1976).

Plan of the Book

The remainder of this chapter will address some definitional and conceptual issues, such as what is meant by *personality, temperament, genes,* and *environment*—but briefly: the aim is to orient the reader, not to pursue philosophical niceties.

The next chapter will introduce some of the important methods of behavior genetics: twins, family studies, adoptions, and various combinations of these, such as studies of the families of identical twins and studies of twins reared apart. We will also look at the model-fitting methods commonly used today to summarize findings and to make inferences in these studies. To keep all of this from being unduly abstract, the various methods will be applied to address a specific question: What are the relative contributions of genes and environment to placing individuals along the personality dimension that runs from the shy introvert at one extreme to the sociable extravert at the other?

Chapter 3 will focus on the genetic and environmental contributions to variation along other important dimensions of personality and temperament. The so-called Big Five dimensions of personality will be used to organize the discussion.

Chapter 4 will ask how the breadth or specificity of trait definition affects behavior genetic analyses.

Chapter 5 will consider changes in personality over time, as assessed both in studies of individuals of different ages and in studies in which the same individuals are measured on more than one occasion. Our concern will be with how the genes and the environment contribute to personality at different ages and to personality change over time.

Finally, chapter 6 will place the results of earlier chapters in a broader context. It will examine such potentially complicating factors as genotype-environment correlation and interaction. It will compare the results for personality with those from neighboring trait domains such as ability. And last, it will attempt to place the findings about individual personality variation in a perspective of evolutionary views about human nature.

Some Basic Concepts

PERSONALITY AND TEMPERAMENT

So far in this chapter, the terms personality and temperament have been used in a broad way to point at the general aspects of behavior with which we will be concerned. We need now to be a little more precise. In general, we will be using these terms to refer to the characteristic manner or style of an individual's behavior, as distinct from the goals toward which it is directed (motivation), or the machinery of its execution (cognitive and motor skills). Temperament and personality may refer to purely stylistic features of behavior, such as the vigor, tempo, or persistence with which it is carried out, or to the emotional expression that accompanies it, such as fearfulness, exuberance, aggressiveness, or self-restraint. The features of behavior so designated must be typical or enduring. Specific episodes of anger at a frustration or grief at a loss do not a temperament make, but characteristic styles of hostile or depressed behavior may be classified as temperament or personality traits. The persistence involved need not be lifelong—otherwise it would be meaningless to speak of personality change—but something must be characteristic of a person over a reasonably extended period of time if it is to be called personality or temperament.

And what is the relationship between personality and temperament? In general we will use personality as the broader term, with temperament restricted to those aspects of personality that are simpler and earlier to emerge in life, often associated with emotional expression. Shyness and reserve with strangers may be characteristic of a 1-year-old's temperamental style. They may be characteristic of a 21-year-old's behavior, too. To the extent that they seem to be the same sort of thing for the latter as for the former, by a more or less direct descent across time, they could be described as temperamental characteristics for the 21-year-old as well. But the 21-year-old's shyness is likely to incorporate additional features, such as an explicit self-consciousness and the traces of a history of previous social encounters, that distance it from its simple roots and lead us to refer to it more broadly as a personality trait.

The reader should be cautioned that there are many different definitions of personality and temperament extant. Allport (1937) listed 50 different definitions of the term personality—and there have been many since. For a discussion of several different contemporary definitions of temperament, see Goldsmith, Buss, Plomin, Rothbart, Thomas, Chess, Hinde, and McCall (1987). One specification sometimes made of temperament—that it is genetically based (e.g., Buss and Plomin, 1975)—will not be made here, because we want to leave ourselves free to ask if indeed it is. In fact, we will discover that personality traits measured in adults often turn out to be more heritable than temperament traits measured in infants—but that is a later, and perhaps a different, story.

GENES, ENVIRONMENT, HERITABILITY

Genes are the information-bearing structures in the DNA (deoxyribonucleic acid) within every bodily cell. They guide the construction of enzymes and proteins and thus act as the blueprints and schedules for development. The environment is the external input to the developmental process, comprising a range of factors from the biochemical (as in cellular nutrition) to elaborately patterned light and sound waves (as in watching *Don Giovanni*). The complete developmental process of any human organism involves an incredibly complex series of interactions between the genes and the sequence of environments within which the organism develops. Earlier parts of this sequence can affect later parts. A given gene need not continually be active—it may be switched on and off in response to activity in other genes or to feedbacks from the immediate biochemical environment (which may in turn reflect more remote environmental inputs).

Given such complexity, one may wonder how anyone could have the temerity to attempt to assess the relative impact of genetic and environmental influences upon any human trait. But fortunately one does not have to understand the developmental process in all its intricate detail in order to make some progress in recognizing the causes of differences among individuals. In the first place, differences in the behavior of quite complex systems can sometimes be due to quite simple causes. The fact that Bill is at the moment a whole lot more active than Tom can be greatly clarified by knowledge of the fact that Bill has just stepped on a thumbtack. In the second place, we may be able to assign causes correctly to a broad category without understanding all the finer details. The fact that Porsches can go faster than Yugos is probably due to design differences between them rather than to the gasoline they use, and the relevant design differences are more likely in the engine and transmission than in the headlights or the brakes.

Heritability refers to the contribution of the genes to individual differences in a particular trait in some particular population. It is a population concept; one does not refer to the heritability of a trait in an individual. The term is used in two senses: a broad sense in which it refers to the total proportion of the variation of the trait that is due to the genes, and a narrow sense, which refers to just that part of the genetic variation that is transmissible across generations. For some genetic effects, genes act individually. These effects are sometimes referred to as *additive*, because the effects of individual genes add together in their effect on the trait. Other genetic effects, known as dominance and epistasis, or collectively, as *nonadditive* genetic effects, depend on the particular configurations of genes that are present: dominance, on the combination of genes present at a given chromosomal locus, and epistasis, on configurations across loci. Because of the reassortment of genes that occurs during the human reproductive process, these latter effects do not get transmitted as such from parent to offspring, although they contribute to the genetic variation within any given generation. Heritability in the broad sense includes both kinds of genetic effects, additive and nonadditive; heritability in the narrow sense includes only the additive kind. An animal breeder trying to change some characteristic in a breed by selection of mating pairs would be most interested in heritability in the narrow sense. A psychologist trying to understand the sources of individual differences would most often be interested in heritability in the broad sense, the total effect of the genes on the trait. Ideally, of course, one would understand heritability in both senses, that is, understand what total proportion of the individual differences in a trait are genetic, and what part of that is additive, and therefore passed on from parent to child. In the next chapter, we will learn how heritability estimates of both kinds can be made for human populations.

CORRELATION, INTERACTION, ERROR

Simple estimates of the relative effects of genes and environment on individual differences in a trait are frequently useful, but a more complete analysis must recognize that genetic or environmental effects on a trait may sometimes be correlated, or may interact. *Gene-environment correlation* refers to the fact that the genetic and environmental influences on a trait may not be independent. For example, emotionally unstable parents may pass on to their children genes predisposing the children to emotional instability and also provide them with environments conducive to the same result. This is a *passive* gene-environment correlation, in the terminology of Plomin, DeFries, and Loehlin (1977), because the correlation occurs independently of the child's behavior. Other types of gene-environment correlation are *reactive* and *active* correlations. In the reactive type, an easily upset child might, for example, be provided with less than normally stressful environments by an understanding teacher (a negative gene-environment correlation) or subjected to extra teasing by classmates on the playground (a positive gene-environment correlation). In each case, the reactions of others to the individual result in a systematic relationship between the genes and the environment relevant to a trait. An active gene-environment correlation results when an individual actively seeks out environments related to the trait in question. Thus, if the child in the preceding example seeks out the teacher and avoids the classmates, he or she would be contributing to an active negative gene-environment correlation.

Gene-environment interaction is different. This term refers to the fact that particular combinations of genes and environments may lead to consequences not predictable from the two considered separately. The phenomenon is thus analogous to an interaction in the analysis of variance—a joint effect different from what would be predicted from the treatments considered independently. Note that we are *not* referring here to the fact that the genes and environment interact extensively during the development of any given individual, but rather to a phenomenon of individual differences: that certain matchings of genes and environments may lead to exceptional outcomes. For example, two genotypes might develop similarly in normal environments, but differ radically in their response to stressful ones. Genotype alone does not predict outcome in this case, nor does environment; the combination is critical.

Finally, *error.* In any empirical study, some of the variation in observed outcome will be due to less than perfect reliability in the measuring instruments or uncontrolled variation in the conditions under which the observations are made. Thus some of the differences among the obtained scores on a set of personality scales administered to a group of individuals will not represent stable or meaningful differences among the individuals at all, but rather will reflect unreliability in the measurement instruments, minor undetected errors in scoring the tests or in entering the data into the computer, and so on. If one wants to assess the genetic and environmental influences on the actual personality traits (as opposed to the fallible scores), one needs to allow for variation due to sources of error of this kind.

In short, if we are attempting to assign observed individual differences to genetic and environmental causes, we must recognize that sometimes the answer can be both (when gene-environment interaction is present), undecided (in the case of gene-environment correlation), or neither (for error). Fortunately, these complications are often fairly minor in their effects, and rarely so severe as to negate the enterprise altogether. Also, as we shall see, there are methods that permit us to estimate and allow for them if necessary.

TRAITS, TYPES, NEEDS, ROLES, ACTS, AND PERSONAL PROJECTS

Psychologists looking at personality have conceptualized it in a variety of different ways at a variety of different levels of generality or specificity. Most personality researchers who have investigated the effects of genes and environment have done so at the level of *traits*—general dimensions of individual differences along which particular individuals may be located. Therefore, most of our attention in this book will be on traits. But one should keep in mind that this is not the only possible option. People might come in distinct subvarieties, or *types* (Jung, 1924; Gangestad & Simpson, 1990). Alternatively, people can be characterized by the extent to which they express different *needs* (Murray, 1938).

Other personality conceptualizations, such as *roles* (Sarbin & Allen, 1968), *acts* (Buss & Craik, 1983), and *personal projects* (Little, 1989) operate at a more specific level, closer to behavior itself. They are thus complementary to broader concepts such as traits, types, or needs. The style or manner in which a person carries out his or her roles, projects, or acts can be seen as a trait; acts, roles, or projects may operate in the service of needs for achievement, for safety, or for love. Most behavior genetic analyses have been carried out at the trait level, and so we will only incidentally be concerned in this book with personality mechanisms of a more specific kind. Traits themselves may be described broadly or narrowly, however, and this is an issue we will confront as we proceed.

Behavior Genetic Methods

Application to Extraversion

In this chapter, we will look at methods that can be used to estimate the relative influence of genes and environment in contributing to individual differences in human characteristics. To keep matters concrete, we will apply these methods to a particular dimension along which humans differ, that of introversion-extraversion.

The terms *extraversion* and *introversion* were introduced by the Swiss psychoanalyst and personality theorist Carl Gustav Jung (1924) to describe two general attitudes that may characterize people: Extraverted persons are primarily oriented toward the external world about them, introverted persons, toward the internal world within them. Jung thought that each person possessed both attitudes, but that one tended to be dominant and conscious, defining the theme of the individual's public life, with the other subordinate and unconscious—being expressed, for example, in dreams and irrational acts.

A number of later writers have adopted Jung's terms extraversion and introversion, but without his underlying hypotheses, to refer to a dimension of personality related to interest, ease, and efficacy in social relationships, a dimension running from the lively, sociable, outgoing extravert at the one extreme to the quiet, shy, reserved introvert at the other. The usual view is that any particular person can be located along this dimension, with most falling somewhere intermediate and only a relatively small proportion lying dramatically out at one or the other extreme. For convenience, I will usually refer to this trait simply as Extraversion, although the full bipolar dimension from introversion to extraversion is meant unless otherwise specified.

This dimension regularly emerges from factor analyses of self-ratings or peer ratings. For example, in recent studies by McCrae and Costa (1985, 1987), individuals rated themselves and were rated by others who knew them well on a series of 80 bipolar adjective scales. In both sets of ratings, a factor emerged that was defined by such adjectives as sociable, fun-loving, affectionate, friendly, spontaneous, and talkative, at the one extreme, and retiring, sober, reserved, aloof, inhibited, and quiet, at the other.

Most multiscale personality inventories include scales tapping this dimension. Some, such as the Eysenck Personality Questionnaire and McCrae and Costa's NEO Personality Inventory have scales measuring Extraversion directly. Others, such as Cattell's 16 Personality Factor Questionnaire, Gough's California Psychological Inventory, Tellegen's Multidimensional Personality Questionnaire, or Jackson's Personality Research Form provide scales that measure traits at a somewhat more specific level; a broad second-order factor resembling Extraversion emerges, however, when these primary scales are factor-analyzed (see Digman, 1990, for a review and references).

Many, including Jung, have supposed an individual's tendency toward introversion or extraversion to be at least partly constitutional.

> As regards the particular disposition, I know not what to say, except that there are clearly individuals who have either a greater readiness and capacity for one way, or for whom it is more congenial to adapt to that way rather than the other. In the last analysis it may well be that physiological causes, inaccessible to our knowledge, play a part in this. (1924, p. 416)

Thus it becomes of interest to ascertain, if we can, the relative roles that the genes and the environment play in determining the differences among individuals in their location on this dimension.

MZ and DZ Twins: Basic Model Fitting

By far the most popular design in behavior genetic studies over the years has been the comparison of monozygotic (MZ, "identical") twins with dizygotic (DZ, "fraternal") twins. Because MZ twins are necessarily of the same sex, most studies have only used same-sex DZ pairs. For the simple estimation of heritability, this is indeed the appropriate comparison. But the comparison of same- and opposite-sex fraternal pairs can provide interesting additional information, and the failure to gather data on opposite-sex DZs in most twin studies is probably unfortunate.

Because MZ twins have the same genes, and DZ twins share on the average only half of their genes, one can take the difference between identical twin resemblance and fraternal twin resemblance on a given trait as reflecting half the effect of the genes on variation in that trait. The correlation between twins is a measure of how much they resemble one another. Thus twice the difference between MZ and DZ correlations on the trait is an estimate of the genetic influence upon it.

There are several nontrivial assumptions being made in using this method of estimating a trait's heritability. First, it is assumed that gene-environment correlation and interaction are negligible for the trait in question. Second, it is assumed that the parents of the twins are not correlated for the trait. (The DZs could average more or less than 50% shared genes if their parents are positively or negatively correlated.) Third, it is assumed that the effects of nonadditive genetic factors such as genetic dominance or epistasis are negligible. (Any nonadditive genetic effects, depending as they do on particular genetic configurations, tend to contribute disproportionately to the resemblance of MZs, in whom genetic configurations, as well as genes, are identical.) Fourth, it is assumed that the MZs and DZs are equally similar in the environmental influences on the trait. This "equal environments" assumption has often worried critics of the twin method. Fifth, and finally, it is assumed that twins are like the other people in the population to which one wishes to generalize. If their personality develops according to an entirely different set of rules, then an estimate based on twins would apply only to twins, and not to people in general.

For the moment, we merely note that these assumptions are being made in this method. Later, we will consider their plausibility for the trait of Extraversion, as well as for personality traits in general. In some cases, such as spouse correlations, direct evidence is available. In other cases, such as that of equal MZ and DZ environments, there has been some empirical evaluation of the assumption. In still other cases, we can compare estimates from methods that do and do not require a given assumption, to see if the estimates differ. If they do not, we conclude that either the assumption is reasonably well met or it is not very critical.

PATH MODELS

Back, then, to the traditional MZ-DZ design. Figure 2.1 expresses the situation in the form of a *path diagram.* In such a diagram, straight, one-headed arrows represent causes, and curved, two-headed arrows represent correlations between the variables they connect. The variables in squares represent observed or *manifest* variables. In this case, T_1 and T_2 represent the scores of a pair of twins on a measure of extraversion. Variables in circles represent unobserved or *latent* variables. G stands for the genetic influences on extraversion, C for the environmental influences common to a set of twins (parents, family, shared friends, etc.), and U for unshared environmental influences, those things that happen to one twin and not the other that might influence the development of extraversion. In the present diagram we are not modeling measurement error separately, so that any shared errors of measurement (for example, due to the twins being tested on the same day) would be included in C, and any random errors of measurement would (by definition) be part of U. The paths, which are labeled by lower-case letters such as *h*, *c*, and *u*, refer to the effects of the genes and shared and unshared environments on the trait in question (technically, they are regression coefficients). For convenient reference, the various symbols used in this and other path diagrams and equations are summarized in a table at the end of the book, before the list of references.

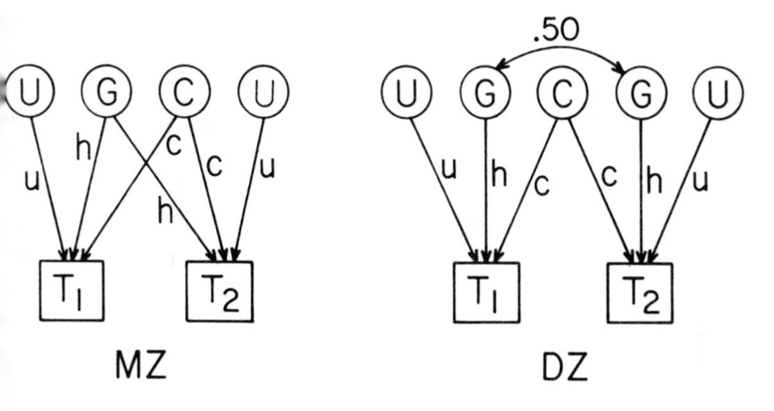

igure 2.1. Path Models of MZ and DZ Twin Correlations
$_1$, T_2 = twins of a pair. G = genotype, C = common environment, U = unshared environment.
aths *h, c, u* = effects of G, C, U on trait.

From a path diagram, assuming standardized variables and follow-ng a few simple rules, one can write an expression for the correlation etween any two variables in the diagram. Essentially, the correlation etween two points is the sum of the paths that connect them, where path is the product of connected arrows leading between the points. he major restriction on such paths is that one cannot go forward and hen afterward go backward along straight arrows on a given path backward first and then forward is allowed). Thus going from U to 3 in Figure 2.1 via *uh* is not legitimate because it involves going orward along *u* and then backward along *h*, but going from T_1 to T_2 ia *hh* is legal, because going backward first and then forward is egitimate. The rule serves the purpose of insuring that variables like J and G will not be correlated simply because they both influence a iven variable, but that variables like T_1 and T_2 will be correlated ecause they share a common cause.

Other rules for defining paths involve avoiding loops and not going hrough more than one curved arrow per path; neither of these is at ssue in the present diagram (for more details, see Loehlin, 1987).

Applying the path rules to the diagram for MZ twins, we can see hat the correlation between the twins can be expressed as the sum of wo paths connecting T_1 and T_2, namely, *hh* and *cc*, or, more compactly,

$r_{MZ} = h^2 + c^2$. For fraternal twins, the correlation is $h\frac{1}{2}h$ and cc, or r_D
$= \frac{1}{2}h^2 + c^2$. We can solve these two expressions algebraically for h^2 a
$2(r_{MZ} - r_{DZ})$, which is the "twice the difference between the MZ an
DZ correlations" mentioned earlier. We can then readily solve for th
variance due to shared environment, c^2, as $r_{MZ} - h^2$.

Does this simple model fit the twin data on Extraversion?

FITTING MODELS TO DATA

Measures of Extraversion have been included in several large twi
studies recently conducted in the United States, Britain, Australia, an
Scandinavia. The obtained correlations for male and female pairs ar
shown in Table 2.1, along with h^2 and c^2 calculated by the simpl
formulas just mentioned. Two points are evident from even a brie
glance at the table. First, the results appear to be fairly regular acros
sexes and populations. And second, they are inconsistent with th
model of Figure 2.1. The shared environmental variance, c^2, as
variance cannot be negative. One or more of the assumptions underlying the analysis must be false for this trait.

Before considering which assumptions may be at fault—and alternative possibilities—let us formalize these impressions of the data b
a procedure which is becoming standard in contemporary behavio
genetic analysis: model fitting. This has become popular for severa
reasons. First, it is a valuable descriptive tool: It provides a convenien
way of combining different pieces of evidence into a single summar
analysis. Second, it allows for statistical inference. Provided certai
assumptions are met, one can perform chi-square tests of whethe
specified models fit the data. And third, given a model that fits, on
can obtain best estimates of the unknown parameters of interest, suc
as the values of h and c in Figure 2.1.

Behavior genetic model fitting is usually done by an iterativ
computer program. (Iterative = repetitive; the program proceed
toward a solution by progressive cycles of trial and error adjustment.
One provides such a computer program with the desired model
either the set of equations or the path model in some appropriat
representation, and the program repeatedly adjusts trial values of th
unknowns (the h^2s and c^2s in this example) until it achieves as good

TABLE 2.1 Extraversion Correlations for MZ and Same-Sex DZ Pairs in Five Large Twin Studies, and h^2 and c^2 Estimates Based on Simple Twin Model

	Britain		*U.S.*		*Sweden*		*Australia*		*Finland*	
	r	*Pairs*	*r*	*Pairs*	*r*	*Pairs*	*r*	*Pairs*	*r*	*Pairs*
Male twins										
MZ pairs	.65	70	.57	197	.47	2274	.50	566	.46	1027
DZ pairs	.25	47	.20	122	.20	3660	.13	351	.15	2304
h^2	.80		.74		.54		.74		.62	
c^2	–.15		–.17		–.07		–.24		–.16	
Female twins										
MZ pairs	.46	233	.62	284	.54	2713	.53	1233	.49	1293
DZ pairs	.18	125	.28	190	.21	4130	.19	751	.14	2520
h^2	.56		.68		.66		.68		.70	
c^2	–.10		–.06		–.12		–.15		–.21	

SOURCES: Britain—Eaves, Eysenck, & Martin (1989); U.S.—Loehlin & Nichols (1976); Sweden—Floderus-Myrhed, Pedersen, & Rasmuson (1980); Australia—Martin & Jardine (1986), correlations calculated from mean squares; Finland—Rose, Koskenvuo, Kaprio, Sarna, & Langinvainio (1988).

fit as possible between the actual observed correlations and the correlations implied by the model.

GOODNESS OF FIT

One might consider using various criteria to assess the goodness of fit of the model to the data—a minimum sum of squared differences, for example. Most current programs use either a weighted least squares or a maximum likelihood criterion. These usually give results similar to simpler criteria such as ordinary least squares, but they weight observations based on the information they provide (giving more weight to those based on larger samples, for example), and they have the important advantage of providing a statistical test: a χ^2 test for the goodness (or rather badness) of fit of the model to the data. A statistically significant χ^2 means that the model in question can be rejected—even with optimum choices for the unknowns the fit to the data is too poor to be attributable to random sampling fluctuation. A nonsignificant χ^2 does not mean that the model is necessarily correct, but at least it is consistent with the data. The degrees of freedom for the χ^2 test are the number of observed values less the number of unknowns involved in the fitting. The χ^2s can be interpreted by referring them to standard tables of the χ^2 distribution. Readers may recall that values of χ^2 increase with the number of degrees of freedom. A useful rule of thumb in model fitting is that a χ^2 roughly equal to its *df* represents a good fit, namely, a p value of near .50, meaning that the chances are 50-50 that one would obtain discrepancies that large simply by chance sampling fluctuation. If χ^2s are larger than their *dfs* this represents poorer fits, until a χ^2 is reached that achieves some conventional level of significance, such as $p < .05$, for rejecting the model in question. A χ^2 less than its *df*, on the other hand, can usually be considered to represent an excellent fit of model to data. For convenience, we will often specify the number of degrees of freedom by a subscript to the χ^2 symbol, thus χ^2_4 would refer to a χ^2 based on 4 *df*. If such a χ^2 were less than the critical value from a χ^2 table required for significance at the .05 level (with 4 *df*, this is 9.49), we would consider the fit to be acceptable, in the sense that the data do not require that the model be rejected; if the χ^2 were 4.0 or less we would consider it to represent a good fit.

There are several differences between this kind of model fitting and the statistical analyses with which readers may be more familiar. What is fitted here are summary statistics, i.e., the correlations in different samples, rather than individual scores. The degrees of freedom for evaluating the fits are based on the number of correlations being fitted, rather than on the number of individuals involved (the latter however, enters into the calculation of the chi-square). Finally, the statistical test is for *lack* of fit; i.e., a large χ^2 and a small p value mean that the model in question does *not* fit the data. Small χ^2s and large p values are what make model fitters happy, in that these values signify consistency of model and data.

Let us illustrate this model-fitting procedure by asking an obvious question: Should the sample-to-sample differences we observe in Table 2.1 be considered real, or are they the sort of thing we might expect just from sampling fluctuation? Let us first look at the males. The procedure will be to take our two equations expressing r_{MZ} and r_{DZ} in terms of h^2 and c^2 and apply them to the data from the five samples simultaneously. The program will vary h^2 and c^2 until it obtains best-fitting values and a χ^2 telling us how well the correlations implied by these values fit the observed MZ and DZ correlations. Since we are fitting 10 observed correlations using 2 unknowns, the χ^2 test will be based on 8 degrees of freedom.

The particular program that was used for the examples in this book is a general-purpose iterative equation-solving program by Davidon (1975), and the statistical strategy followed is one suggested by Rao, Morton, Elston, and Yee (1977), in which both implied and observed *r*s are transformed to Fisher's *z*s to improve normality and yield some economies of computation. I should emphasize that there is nothing magic about this particular program and procedure—I have used it because it is convenient and can be applied readily in various cases we will be considering later. More and less elaborate model-fitting procedures exist. At the elaborate end of the range are programs that fit directly to the raw data in irregular family pedigrees (e.g., Lange, Westlake, & Spence, 1976). At the other extreme, simple twin path models can be solved by any of a variety of widely used iterative model-fitting programs, for example, EQS (Bentler, 1989) or LISREL (Jöreskog & Sörbom, 1989). In fact, a recent issue of the journal *Behavior Genetics* was devoted entirely to behavior genetic model fitting using LISREL (Boomsma, Martin, & Neale, 1989).

RESULTS FOR THE TWIN DATA

A number of model-fitting runs to the data of Table 2.1 are summarized in Table 2.2.

The first row of the table answers our initial question about consistency—it shows the result of fitting the equations to the male twin correlations in the five studies. This result is an overall χ^2 of 13.68, which with 8 *df* is not statistically significant at the .05 level. Thus the sample-to-sample differences among the male pairs in Table 2.1 are indeed of a sort that might reasonably have resulted from sampling fluctuation. The best-fit overall estimates for h^2 and c^2 are .60 and –.12, respectively; they can be considered as a kind of weighted average of the figures arrived at in the separate studies in Table 2.1.

TABLE 2.2 Some Models Fit to MZ and DZ Twin Correlations for Extraversion

Question	*Answer*	χ^2	*df*	*p*
1. Does the model fit the male data?	Yes	13.68	8	> .05
2. Does the model fit the female data?	No	21.42	8	< .01
3. Could both sexes have the same h^2 and c^2?	No	10.18	2	< .01
4. Could their h^2s be the same?	Yes	3.25	1	> .05
5. Could their c^2s be the same?	Yes	.82	1	> .30
6. Could the c^2s be zero?	No !	53.12	1	<< .001

NOTE: Tests in lines 3-6 are tests of χ^2 differences. See text for details.

What about the female pairs? In this case (next row) the result is a χ^2_8 of 21.42, which *is* statistically significant ($p < .01$). That is, there do appear to be some dependable sample-to-sample differences for the female pairs. The fluctuations among the correlations in Table 2.1 are not conspicuously greater for the females than the males, but the sample sizes are larger for the females, allowing us to conclude with some confidence that at least some of the differences among the samples are real.

Actually, it is not very surprising that there should be some differences across the samples. Although all of the studies used self-report questionnaire measures of Extraversion based on Eysenck's conception of the trait, the actual scales differed. The Swedish and Finnish studies used translations of a short form of an Eysenck scale, the U.S. study used an a priori scale by Eysenck scored on the items of the California Psychological Inventory, and the British and Australian studies used the regular Extraversion scale from the Eysenck Personality Questionnaire (EPQ). The samples were differently selected. The British and Australian studies were based on volunteer adult twin samples recruited via advertising and other publicity, the U.S. sample was identified from among high school juniors who took the National Merit Scholarship Qualifying Test in 1962, and the Scandinavian studies were based on a follow-up from population birth records. Ages differed. The U.S. sample were mostly 17- and 18-year-olds when tested, the other samples covered a range of adult ages. Extraversion scores tend to decrease slightly with age: The correlation with age was about –.15 in the Australian sample. Within the Swedish sample, twin correlations tended to be somewhat higher in the younger birth cohorts. The British data were adjusted for age, the others were not.

Further model fitting, not shown, suggests that the Finnish sample, which has somewhat lower correlations and large *N*s, could be the primary source of the inconsistency for females. Without it, the χ^2 with 6 *df* is 9.62, which does not represent a statistically significant departure from fit ($p > .10$). Nevertheless, in view of the fact that the Finnish data are not grossly discrepant for females and are statistically consistent for males, and that there is no a priori reason for considering the Finnish study anomalous (methodologically, it was very similar to the Swedish study), we will retain the Finnish data for the present, although we will exclude them in some subsequent analyses. In short, although the female samples do differ among themselves, and these differences, because of the large samples involved, are probably not entirely due to chance, the consistency across studies is more impressive for both males and females than are the inconsistencies.

χ^2 DIFFERENCE TESTS

A further attractive feature of the χ^2 goodness-of-fit test is that it can easily be used to test the statistical significance of the *difference* between the fit of two models, provided that one is nested within the other, i.e., that they are the same except that one has more unknowns. The test is done simply by taking the difference between the two χ^2s as a χ^2 and the difference between the two *df*s as a *df*.

Let us illustrate this point by looking at sex differences. In the analysis for females, the best-fit values across all five female samples are $h^2 = .68$ and $c^2 = -.16$. Do these differ significantly from the best-fit values of $h^2 = .60$ and $c^2 = -.12$ obtained for the males? We can test this difference by a difference in χ^2. I suggest that you follow the next few paragraphs step by step with pencil or calculator in hand. Once you get a feel for such analyses, you should be able to deal easily with a number of similar tables to follow.

The χ^2 for fitting a single value of h^2 and a single value of c^2 across all 10 samples in Table 2.1 turns out to be 45.28, with 20 observed correlations minus 2 unknowns = 18 *df*. The χ^2 if separate h^2s and c^2s are fitted in the male and female subsamples is 35.10 with 16 *df*. The models are nested, because the second model is the same as the first, except that we have added a couple of additional unknowns in solving for h^2 and c^2 in the two sexes separately. The difference of 35.10 from 45.28 provides the difference χ^2 of 10.18 shown in Line 3 of Table 2.2, and the difference between the *df*s of 16 and 18 is its *df*. Consulting an ordinary table of χ^2, 10.18 with 2 *df* proves to be statistically significant ($p < .01$). Thus we can conclude that the results of our model fitting differ for the two sexes (in addition to differing across studies in the case of the females).

Is the sex difference in the h^2 or in the c^2 or both? Again, model-fitting tactics and differences in χ^2 can be employed. To test for a difference in h^2, we can fit a model in which h^2 is required to be the same for males and females, but c^2 is allowed to differ, and compare its χ^2 to the 35.10 above. For a test of c^2, we do the same thing, but require the c^2s to be the same.

The test with h^2s constrained to be the same yields a χ^2_{17} of 38.35. The difference from 35.10 is equal to a χ^2 of 3.25 (Line 4 of Table 2.2), which with 1 *df* is not statistically significant ($p > .05$). The other test, requiring the c^2s to be the same, yields a χ^2_{17} of 35.92. The difference χ^2 in this case is .82 (Line 5), which is also well short of statistical significance, with 1 *df* ($p > .30$). Thus we can conclude that there is a sex difference somewhere, but either a difference in h^2 or c^2 is consistent with the data.

We can use the same method to ask if the c^2s really are negative, and not just chance departures from zero. Of course, in this case it is hardly necessary to carry out a statistical test. Would an effect that occurs in each of 10 separate samples be due to chance? Perhaps the chief merit of actually performing the test is to find out if it agrees with our common sense. We do this by setting c^2 to zero and seeing whether this leads to a significant increase in χ^2 over the 45.28 obtained when h^2 and c^2 are required to be the same across samples, but c^2 is allowed to be negative. The resulting χ^2_{19} is 98.40, and the difference, a χ^2 of 53.12 with 1 *df* (Line 6), is highly unlikely to be a result of chance (with 1 *df*, a χ^2 of 10.83 would yield statistical significance at the .001 level).

Yes, those c^2 estimates really are negative, and thus this model will not do for Extraversion.

Alternative Models and Assumptions for Twin Data

Let us look at some alternatives. For convenience, we will do this just with the male samples, to avoid having to concern ourselves with sex and study differences at this stage.

CONTRAST EFFECTS

One of the assumptions made in the previous model (Figure 2.1) is that there are shared environmental influences for the twins due to their having the same parents, being the same age, having the same friends, and so on, that affect their Extraversion in the same way. Suppose that this is emphasizing the wrong point. Suppose that what the parents and friends do (that matters) is that they go around saying, "Larry is the extravert and Barry is the introvert." We can represent this state of affairs by Figure 2.2, which is just like Figure 2.1 except that the two *c*s have opposite signs. Now the equations are $r_{MZ} = h^2 - c^2$ and $r_{DZ} = ½h^2 - c^2$. The model fitting with these equations can be carried out just as in the preceding case. The χ^2 is 13.68, as before, and the *df* remains the same, as do the estimates of h^2 and c^2, except that the value of c^2 is now positive, as befits a variance. In this view, the shared environment of twins does have an effect on the extraversion-introversion dimension, but it is a contrast effect—it works to push the two twins of a pair in opposite directions. Later, we will able to examine the merits of this model with other data; such a contrast effect should be absent if, for example, twins are reared apart. In the meantime, let us consider some other possibilities.

NONADDITIVE GENETIC VARIANCE

Suppose that the failure of the original model lies in the failure of a genetic rather than an environmental assumption. Perhaps we were

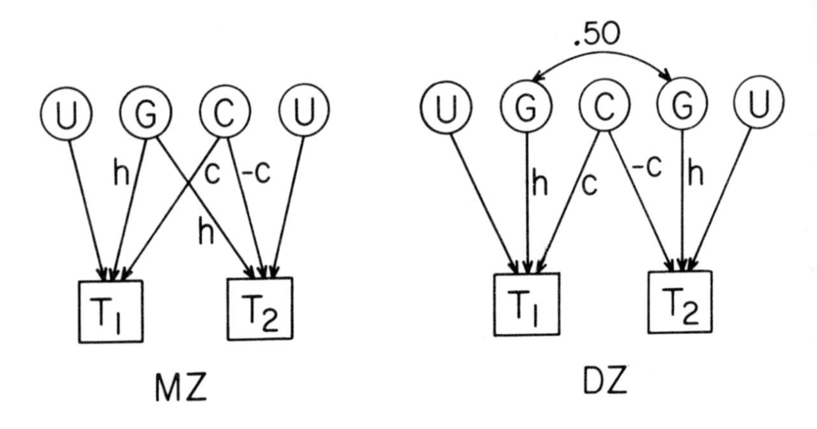

Figure 2.2. Alternative MZ-DZ Twin Model: Contrast Effects

T_1, T_2 = twins of a pair. G = genotype, C = common environment, U = unshared environment. Paths h, c = effects of G, C on trait. (To help the reader in tracing paths, in this and subsequent figures only arrows which enter into equations are labeled.)

wrong to assume that genetic dominance and epistasis are negligible for Extraversion.

Figure 2.3 represents a model that assumes the effects of shared environment are absent, but that nonadditive genetic variance is present. Two different versions of the model are shown, one at the top which assumes that all the nonadditive genetic variance is due to genetic dominance, and one at the bottom which assumes that all of it is due to epistasis involving interactions among many genes, the situation that Lykken (1982) has referred to as *emergenesis*. In the former case, genetic theory predicts that the nonadditive component of the trait will be correlated .25 for DZ twins; in the latter case, the nonadditive contribution to the DZ correlation will be zero, as there will be a negligible probability that the two DZ twins will simultaneously inherit the same configuration if many separate genes are involved. Intermediate hypotheses—for example, that effects of both kinds are present—will yield results intermediate to these. In all the variants of this hypothesis, MZ twins, being genetically identical, will share configurational genetic effects completely as well as additive effects. The equations for the cases of pure dominance and pure epistasis are shown in Table 2.3. The fits in each case have the same

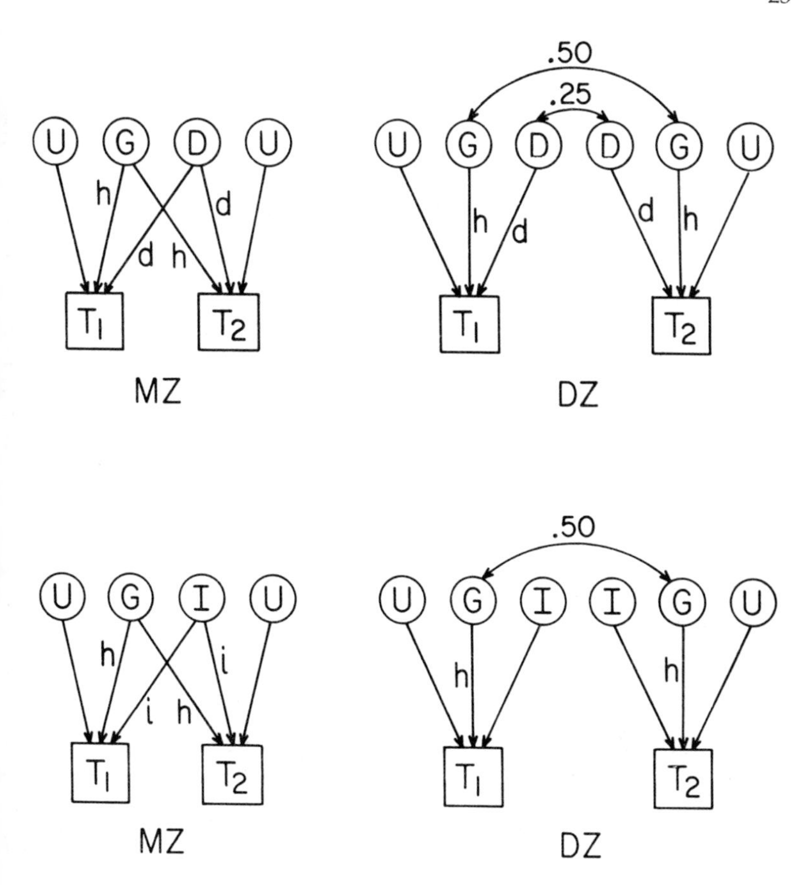

Figure 2.3. Alternative MZ-DZ Twin Models: Nonadditive Genetic Variance
Top: Genetic dominance. Bottom: Multiple-gene epistasis. T_1, T_2 = twins of a pair. G = additive genotype, D = genetic dominance, I = multiple-gene epistasis, U = unshared environment. Paths h, d, i = effects of G, D, I on trait.

χ^2, 13.68, with 8 *df*. For the pure dominance case, $h^2 = .24$ and $d^2 = .24$. For the extreme epistasis case, $h^2 = .36$ and $i^2 = .12$. The broad-sense heritability is the same in both cases, i.e., .48, but the proportion of it that is estimated to be additive differs.

As one might suspect, the fact that the same χ^2s are obtained for the goodness of fit to the data of these and the preceding models is no accident. The solved-for values in each instance are simply different

TABLE 2.3 Equations for MZ and DZ Twins, Nonadditive Genetic Variance

Only dominance

$$r_{MZ} = h^2 + d^2$$
$$r_{DZ} = \tfrac{1}{2} h^2 + \tfrac{1}{4} d^2$$

Only multiple-gene epistasis

$$r_{MZ} = h^2 + i^2$$
$$r_{DZ} = \tfrac{1}{2} h^2$$

NOTE: Correlations are for MZ and DZ twins; path symbols as in Figure 2.3.

algebraic functions of the same MZ and DZ correlations. The χ^2s are reflecting the consistency of these correlations across the five samples. What differs are the numerical estimates of h^2, c^2, d^2, or i^2 that these correlations imply under the different assumptions incorporated in the models.

One can confirm that the inconsistency reflected by the χ^2 is among the correlations themselves, rather than among particular interpretations of them, by fitting to the data a model in which the unknowns are just the population MZ and DZ correlations; i.e., one asks what values for the correlations can be found that best fit the data from all five samples. The χ^2_8 from such a test of homogeneity across the male samples is the familiar 13.68 (for the females, 21.42). The overall best estimates of the MZ and DZ correlations are .48 and .18 for the male twins and .53 and .19 for the female twins. (One should not assume from this example that tests of homogeneity of correlations and tests of the fits of particular models will always yield the same chi-square—for the more complex models we will consider later, this will not be the case.)

ASSORTATIVE MATING

One assumption made in the models so far is that the additive genetic correlation between DZ twins is .50. This is in turn based on the assumption that spouses are not matched to one another on the trait. Is this plausible for Extraversion? In two British samples totaling 889 pairs, a pooled correlation between spouses of .065 was reported for the Extraversion scale of the EPQ (Eaves et al., 1989). In a study in

Hawaii, 422 spouses were correlated .12 on the Introversion versus Extraversion scale of an earlier Eysenck questionnaire, the Eysenck Personality Inventory (Ahern, Johnson, Wilson, McClearn, & Vandenberg, 1982). In a sample of adoptive families in Texas (described in Loehlin, Willerman, & Horn, 1985), the correlation between spouses on the Eysenck Extraversion scale of the CPI is .057, for 189 pairs. Spouse correlations this small, under typical assumptions, should produce only a trifling departure of the genetic correlation between siblings from .50, and thus present no difficulties for twin models of the kind we have been discussing.

UNEQUAL ENVIRONMENTS

Suppose that environmental resemblance really does have a positive, not a negative effect on twin resemblance (i.e., the contrast model of Figure 2.2 does not hold), but the equal environments hypothesis is false. Figure 2.4 shows the path model in this situation with different values of c allowed for MZ and DZ pairs. Without further data, such a model has too many unknowns to be solved, but we can look at extreme cases. In the one, we will let c^2_{DZ} take on the minimum value it can under this general hypothesis, namely zero, and in the other, we will let h^2 take on its minimum value, also zero.

The fit of the model to the male twin data in either case remains the same $\chi^2_8 = 13.68$. With c^2_{DZ} set to zero, $h^2 = .36$ and $c^2_{MZ} = .12$. At the other extreme, with h^2 set equal to zero, c^2_{MZ} and c^2_{DZ} are simply equal to the fitted MZ and DZ correlations, i.e., .48 and .18.

Thus, if the assumption of equal resemblance of MZ and DZ twin environments is false, we have still another means of fitting these data while avoiding negative variance estimates.

THE EQUAL ENVIRONMENTS ASSUMPTION

The assumption that trait-relevant environments are equally similar for MZ and DZ twin pairs is often referred to as the "equal environments" assumption. This assumption has received a fair amount of direct investigation, so we can ask how plausible it is empirically. For example, in the National Merit twin study (Loehlin

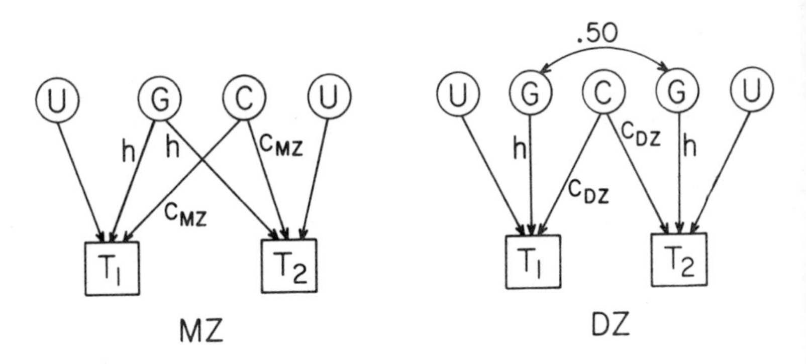

Figure 2.4. Alternative MZ-DZ Twin Models: Unequal Shared Environmental Effects

T_1, T_2 = twins of a pair. G = additive genotype, C = common environment, U = unshared environment. Paths h, c_{MZ}, c_{DZ} = effects of genotype and MZ and DZ shared environments on trait.

& Nichols, 1976) this assumption was examined in the following way. Parents in the study rated a number of different kinds of environmental similarity concerning their (MZ or DZ) twin children: whether the twins had been dressed alike, how much they had played together as children, how much time they had spent together as adolescents, whether they had usually had the same teachers in school, whether they had usually slept in the same or separate rooms, and whether the parents in raising them had tried to treat them alike or differently. On the average, the parents of DZ twins indicated more differences than did the parents of MZ twins. But we still need to be concerned with the direction of cause and effect. Are DZ twins more different because they are treated more differently, or are they treated more differently because they are more different to start with? A way of resolving this issue is to look within the group of MZ twins. Some identical twins are dressed alike, some are not. Some are nearly always together, others are not. Some parents try to treat twins alike, others try to treat them differently. If differences on environmental variables such as these are responsible for the greater personality resemblance of MZ than DZ twins, we would expect them also to affect personality resemblance within the MZ group. MZs dressed alike should be more similar than MZs not dressed alike, and so forth. (This should occur within the DZ group, too, but the MZ group is a simpler case because there are no within-pair genetic differences to complicate the picture.)

In fact, in the National Merit data there was a slight tendency toward such a correlation between treatment differences and personality differences among MZs, but it was very slight. For example, among 108 correlations of treatment differences and differences on the scales of the CPI (6 ratings × 18 scales), there were 74 (69%) that were positive. But all were small—the total range of correlations was from –.08 to +.13, and the median value was +.02. Considering only the three CPI scales most closely related to Extraversion (Dominance, Sociability, and Social Presence), the corresponding figures were 17 out of 18 correlations positive, a range from –.01 to +.11, and a median of +.04. Thus, the fact that MZ twins are more often dressed alike, play together, and so forth, may mean that their effective environments with respect to Extraversion are a little more alike than those of DZ twins, but not much. (There could even have been no difference, because there is an alternative possibility: The current similarity of the twins might have influenced their parents' retrospective judgment of the similarity of treatment that they had received.) At best, the size of the effect is not such as to suggest that much of the MZs' greater resemblance in Extraversion can be attributed to it.

Another variable that has been studied is looking alike. MZ twins look more alike than DZ twins. Could this circumstance lead to their greater similarity in personality? (This can be considered a case of a reactive gene-environment correlation, with the mediating variable being the responses of others to the twins.) The test is the same: Some MZ twins look more alike than others do; are the ones that look more alike more similar in Extraversion than the ones who look less alike? In a couple of studies of young MZ twins the answer was no. In two separate samples in one study (Plomin, Willerman, & Loehlin, 1976), the correlations were .02 and –.03 between parent-rated similarity of the twins' appearance and similarity of their sociability. (There were 60 pairs in each sample.) In another study of 51 young MZ pairs, the correlation between similarity of appearance and similarity of scores on an Extraversion factor from the Children's Personality Questionnaire was –.05 (Matheny, Wilson, & Dolan, 1976). In a third study of college-age twins, physical attractiveness was associated with Extraversion, substantially so among males, using an Extraversion scale from the Comrey Personality Inventory and independent judges' ratings of physical attractiveness (Rowe, Clapp, & Wallis, 1987). The authors' adjustments for resemblance in physical attractiveness had a negligible effect on the twin correlation for Extraversion, however, suggesting that the equal environments assumption for this trait is probably not compromised in any major way by differences in physical resemblance between the two kinds of twins.

Another variable that has been investigated relative to the assumption of equal environmental resemblance is the amount of contact between adult twins. In the large Swedish study (Floderus-Myrhed, Pedersen, & Rasmuson, 1980), twins were asked how frequently they had contact with each other. About half the MZ pairs reported daily or almost daily contact. Only about one-third of the DZ pairs did. Does frequency of contact go with greater resemblance in Extraversion among MZ pairs? Yes, but only slightly so: The correlations were .08 and .11 for male and female pairs, respectively. A similar finding was reported for Extraversion in the Finnish twin study, but only for female pairs (Rose & Kaprio, 1988), and neither study showed much relationship within the DZ group.

On the whole, then, it seems unlikely that much of the greater resemblance of MZ than DZ twins for Extraversion is due to the greater similarity of the environments to which the MZs have been exposed, although a small effect of this kind cannot be ruled out.

ARE TWINS LIKE NONTWINS?

The relevance of any of the analyses described above to the question of how genes and environment contribute to personality variation in the general population depends on twins being enough like nontwins to permit us to extend to the general population conclusions that are arrived at from twin studies. One part of the answer depends on the consistency of results from behavior genetic studies that involve twins with those of other behavior genetic studies that do not, such as those based on adoptions. But there also have been some direct comparisons. For example, in the National Merit study, Nichols and I compared the twin sample with a control group of nontwin individuals also selected from among National Merit Scholarship test takers and measured with a similar set of questionnaires. On 131 different personality, attitude, and behavioral variables, we found few differences between the groups. The twins appeared on the average to be quite similar in personality to similarly chosen nontwins.

A NOTE ON CORRELATIONS VERSUS OTHER STATISTICS

In this chapter, we have been fitting our models to twin correlations. Others (e.g., Cattell, 1982; Eaves et al., 1989) prefer to fit their models to the between- and within-pair mean squares from an analysis of variance based on the twin data. In practice, these methods usually lead to similar conclusions. With twin data one can readily transform mean squares to correlations via the relationship $r = (\text{BMS} - \text{WMS})/(\text{BMS} + \text{WMS})$, where BMS and WMS refer to between- and within-pair mean square, respectively. The reader interested in a closer examination of model fitting via mean squares will find examples in the book by Eaves and his colleagues (1989) in which such methods are applied to data from a number of the samples described in the present chapter (see especially their chapters 5 and 6). Behavior genetic models can be fit to means as well as to correlations or mean squares (McArdle, 1986). This has been done only rarely with personality variables, but an example will be mentioned in a later chapter.

One should not take the emphasis on chi-squares and model fitting in this book to mean that no other methods are used in behavior genetic analyses. For example, an elegant approach has been developed by DeFries and Fulker (1985), which estimates heritability by using multiple regression to predict one twin's score from that of the other twin. This method is particularly valuable when analyses are made using extreme groups, as when one twin is selected for the presence of some form of psychopathology.

TWIN STUDIES ARE NOT ENOUGH

I hope it is obvious by now that although relevant to the issue of genetic and environmental influences on a trait, twin studies have limitations. Given the assumptions embodied in a given model, they permit us to estimate genetic and environmental influences, but they do not necessarily allow us to decide among the assumptions. We were able to conclude that one popular model, the simple twin model of Figure 2.1, could be rejected for the trait of Extraversion because it led to negative variance estimates, but the data were consistent with several other models involving quite different assumptions. For some assumptions, such as the correlation among spouses and equal twin environments, there was a certain amount of direct evidence available. Other matters remain ambiguous. We need to look at different sources of data if we hope to resolve these questions.

Adoptions and Other Twin and Family Designs

ADOPTIONS

After the twin study, the next most popular behavior genetic design is the adoption study. It is well known that in ordinary families there is some (albeit fairly modest) resemblance of family members on personality traits (evidence summarized in Loehlin, Horn, & Willerman, 1981, Table 11). It becomes of some interest, then, to look at adoptive families. Here we find parents with genetically unrelated children reared as offspring and pairs of genetically unrelated children growing up together as siblings. If parent-child or sibling resemblance persists in adoptive families, we must attribute it to family environmental effects—imitation of the parents by the children; direct communication of values by parents to children or among siblings; the influence on family members of shared environmental factors such as the social or economic status of the neighborhood, television programs jointly watched, books and magazines in the home, diet, whatever. If, on the contrary, personality resemblance vanishes when we examine adoptive families, it will suggest that the genes cause this resemblance in ordinary families. For parent-offspring comparisons, we can rule out the effects of genetic dominance and epistasis in causing resemblance; only additive genetic factors are implicated.

Nor, presumably, would the special environmental circumstances applying to the twin situation be at issue here. A general point worth emphasizing is that any given behavior genetic method involves assumptions, but insofar as these differ from method to method, conclusions that hold up across methods will carry increased weight.

Table 2.4 gives correlations for Extraversion for various relationships from three studies involving adoptive families, carried out in Great Britain (Eaves et al., 1989, p. 127), Minnesota (Scarr, Webber, Weinberg, & Wittig, 1981), and Texas (Loehlin et al., 1985). All of the offspring generation in these studies were old enough to be measured by the same questionnaire as their parents (≥14 years in the Texas study, 16-22 years in the Minnesota study, and adult in the British study). Most of the children had been placed with their adoptive families early in life—91% within the first 6 months in the Minnesota study, almost all within their first few days in the Texas study—so they would have had ample exposure to the adoptive family environment. The biological relationships were taken from comparable non-adoptive families in the British and Minnesota studies, and from the adoptive families themselves in the Texas study, in which they were based on biological children born to the adoptive parents. The sample sizes in the adoption studies are considerably smaller than those in the twin studies discussed earlier. There were 115 adoptive families in the Minnesota study, 150 in the British study, and 220 in the Texas study. The numbers given in Table 2.4 refer to the numbers of pairings on which the correlation is based. A particular individual may enter more than once into such pairings—for example, a father with two adopted children would be paired separately with each in calculating the father-adopted-child correlation, and the two children would enter into the adoptive sibling correlation as well. The resulting lack of independence can be expected to deflate the χ^2s a bit, but should not much affect the estimates of the values of the unknowns (McGue, Wette, & Rao, 1984). The sample sizes vary for the different relationships, but all correlations are based on at least 50 pairings, except for the very small number of biological sibling pairs in the Texas study—there were not many families with two or more biological children in addition to the adopted child that brought the family into the study. The Extraversion scale was from the EPI in the Minnesota study and

TABLE 2.4 Extraversion Correlations in Three Adoption Studies

	Britain		*Minnesota*		*Texas*	
	r	*Pairs*	*r*	*Pairs*[a]	*r*	*Pairs*
Mother and biological child	.21	309	.04	255	–.03	57
Father and biological child	.21	236	.21	255	.20	56
Mother and adopted child	–.02	127	–.03	187	.00	257
Father and adopted child	–.03	93	.05	182	.03	247
Biologically related children	.25	418	.06	135	.13	17
Adoptively related children	–.11	58	.07	75	–.13	125

SOURCES: Britain—Eaves, Eysenck, & Martin (1989); Minnesota—Scarr, Webber, Weinberg, & Wittig (1981), data from unpublished appendix, courtesy the author; Texas—Loehlin, Willerman, & Horn (1985), unpublished data.
a. Estimated.

from the EPQ in the British study; the Texas study used the Eysenck a priori Extraversion scale from the CPI.

Some of the individual correlations look a little odd (e.g., the low biological mother-child correlations in Minnesota and Texas). With sample sizes such as these, a fair amount of sampling fluctuation in the correlations is expected. As a preliminary step to model fitting, we can ask if the observed correlations across the three adoptive samples are consistent within sampling error. We do this by fitting a model that takes as unknowns the population correlations; i.e., we carry out a test of the homogeneity of corresponding correlations across samples, similar to that performed earlier with the twin data. This model has a χ^2_{12} of 11.84, $p > .30$. Thus the hypothesis is quite tenable that the observed correlations are consistent across the three samples apart from sampling error.

PATH MODELS OF BIOLOGICAL AND ADOPTIVE FAMILIES

Figure 2.5 shows path models representing biological and adoptive families, each with two children. The diagrams are alike in their environmental portions, but the biological family also has genetic connections among its members. The squares represent observed Extraversion scores for mothers, fathers, and two children in each kind of family. The circles represent three latent variables for

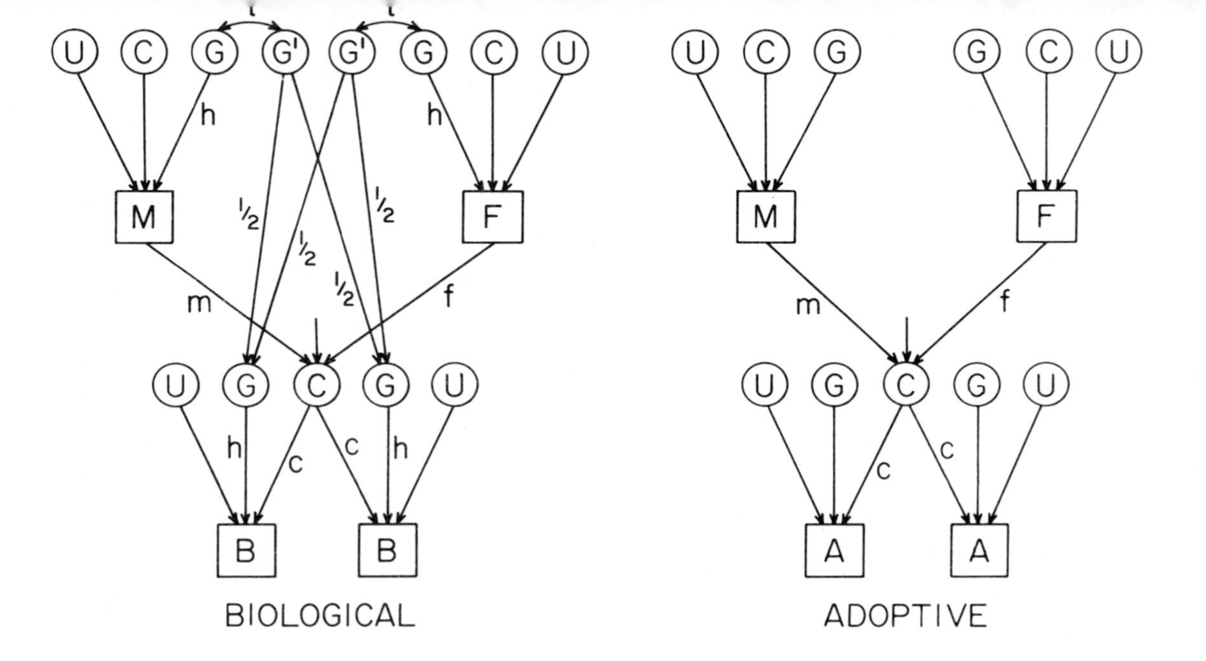

Figure 2.5. Path Model of Biological Family (Left) and Adoptive Family (Right)

M, F = mother, father, B, A = biological, adoptive child. G = genotype, C = common environment, U = unshared environment, G′ = parent's genotype in childhood. Paths *h, c* = effects of G, C on trait, *m, f* = effects of mother's and father's trait on shared environment of children, *t* = correlation between childhood and adult genotype.

each individual: G for additive genetic influences on the trait, C for common or shared family environment, and U for everything else, including environmental factors unique to the individual, gene-environment interaction, and random measurement errors. The paths *h* and *c* have the same meanings they had in the twin design; the paths *m* and *f* refer to the influence of a parent's trait on the environment affecting his or her child's trait. The extra unlabeled arrow pointing to C means that children's shared environment is not solely determined by parental traits—other factors are assumed also to affect it. It will be noticed that the genetic part at the top of the diagram in the biological families involves *two* genotypes, connected with a correlation arrow *t*. This is to allow for the possibility that the genes affecting a given trait may differ at different ages. The symbol with a prime refers to the parent's genotype at the age at which the children are being measured, the ordinary symbol to the parent's present genotype. A low degree of gene-based parent-offspring resemblance might be due to a low cross-age genetic correlation, *t*, between the ages at which they are measured, a low influence of genes on the trait, *h*, or both. (In principle, we could consider different *h*s at the different ages as well.)

The absence of curved arrows connecting the mother's and father's sides of the diagram means that we are assuming that spouses are essentially uncorrelated for the trait of Extraversion, an assumption that as we have seen is reasonably well supported by the facts. The path model on the right in the figure also assumes that there is no systematic resemblance in personality between the birth and adoptive parents of the adopted child. If there were, there would need to be additional paths in the right-hand diagram to reflect this. If adoption agencies tried to place infants selectively in adoptive homes by matching adoptive and birth parents on personality, correlations of this kind could arise. Such a matching process has been reported in adoption studies of IQ. Fortunately, for our purposes, selective placement appears to be negligible for personality traits (Loehlin, Willerman, & Horn, 1982; Plomin & DeFries, 1985).

Table 2.5 gives the equations for various family correlations as derived from the path diagrams. The path tracing in these more complex two-generational diagrams is not quite as simple as with

TABLE 2.5 Equations for Adoption Studies

$$r_{MB} = \tfrac{1}{2}\,th^2 + mc$$
$$r_{FB} = \tfrac{1}{2}\,th^2 + fc$$
$$r_{MA} = mc$$
$$r_{FA} = fc$$
$$r_{BB} = \tfrac{1}{2}\,h^2 + c^2 + \tfrac{1}{2}\,cth^2(m+f)$$
$$r_{AA} = r_{BA} = c^2$$

NOTE: Correlations r_{MB}, etc., for groups listed in Table 2.4; path symbols as in Figure 2.5. r_{BA} = correlation between biological and adopted child in family with both.

twins, but the same basic rules apply. For a fairly easy example, the first correlation, between a mother and either one of her biological children, consists of the two paths (starting from M) of $ht\tfrac{1}{2}h$ and mc. For a harder one, r_{BB} consists of five paths: two with the value $h\tfrac{1}{2}\tfrac{1}{2}h$, one via each parent, the path cc, and two paths $cmht\tfrac{1}{2}h$ and $cfht\tfrac{1}{2}h$, yielding the expression given in the table.

Proceeding to fit 5 unknowns, h, c, m, f, and t, the same in each of the three samples, yields a χ^2_{13} of 13.33 (Table 2.6, Line 2), an acceptable overall fit, although it involves some implausible parameter estimates (the t, m, and f parameters are not well resolved). But a simple two-parameter solution solving for just h and c (setting m and f to zero and r to 1.0) does well enough, with a χ^2_{16} of 16.57 (Line 3). The difference, a χ^2_3 of 3.24, is not close to statistically significant ($p > .30$). The values of the unknowns for this solution are $h^2 = .35$ and $c^2 = 0$.

Looking back at the various twin models we fitted, this result would be most consistent with the results from the model of additive and epistatic genetic variance, which assumed a c^2 of zero and yielded an h^2 of .36 for the males (for the two sexes jointly, the estimate is .366, with $i^2 = .142$).

However, there are more designs and data to look at before we draw any final conclusions.

TWIN-FAMILY DESIGNS

A recent design that has occasionally been used by behavior geneticists is based on twins and other members of their families. The particular variant that we will consider here involves adult MZ twins,

TABLE 2.6 Some Models Fit to Adoption Study Correlations for Extraversion

Question	*Answer*	χ^2	*df*	*p*
1. Are the *r*s from the three studies consistent?	Yes	11.84	12	> .30
2. Can a 5-unknown model fit across samples?	Yes[a]	13.33	13	> .30
3. Would it, if only *h* and *c* were solved for?	Yes	16.57	16	> .30

a. Some paths have unreasonable values, however.

each of whom is married and has children. This situation is interesting because, genetically, an MZ twin just as much resembles his or her twin's children as his or her own. Thus we have the same genetic parent-offspring resemblance occurring in the same family (an MZ twin and his or her own children) and across different families (an MZ twin and his or her twin's children). In addition, the children of the two families, although cousins socially, are related genetically as half-siblings (i.e., as would be the children of the same individual with two different spouses).

Data from two relevant twin-family studies have been reported and are shown in Table 2.7. One, done in Sweden (Price, Vandenberg, Iyer, & Williams, 1982), administered two Extraversion scales—the data from Eysenck's EPQ scale are used in Table 2.7. The study of U.S. veterans (see Loehlin, 1986b) used the Thurstone Temperament Schedule, an inventory that lacks an Extraversion scale as such. Three of its scales, Sociable, Dominant, and Impulsive, go together on a second-order factor which appears to represent Extraversion, however. The values given in Table 2.7 are correlations averaged across these three scales (via Fisher's *z*-transformation). Because of its origin in a sample of World War II veterans, all of the MZ pairs in this study are male; the Swedish sample includes both male and female pairs. The original Swedish study included DZ twins and their families as well. Only correlations corresponding to ones available from the U.S. study have been included in Table 2.7, pooling across sexes as necessary. Ordinary parent-offspring and sibling correlations were not reported separately for MZ and DZ families in the Swedish data and thus are left combined here. This should not make a difference for the models to be tested.

TABLE 2.7 Extraversion Correlations in Two Twin-Family Studies

	Sweden		U.S.	
	r	Pairs	*r*	Pairs
MZ twins	.49	72	.34	44
Twin and own child	.25	264[a]	.18	149
Twin and MZ twin's child	.15	71	.24	121
Siblings	.33	75[a]	.15	102
Cousins via MZ twins	.13	54	.18	84

SOURCES: Sweden—Price, Vandenberg, Iyer, & Williams (1982); U.S.—see Loehlin (1986b).
a. Includes some in families of DZ twins.

An initial test of the homogeneity of corresponding correlations across the two studies by model fitting suggested that the assumption of consistency could reasonably be made ($\chi^2_5 = 3.35, p > .50$).

A path diagram for the twin-family design is shown in Figure 2.6. Two married male MZ twins are shown as T_1 and T_2, along with their respective children, the Bs at the bottom of the figure. Nonadditive genetic variance is represented as epistatic. For simplicity, the twins' spouses are omitted from the diagram—they enter only into the derivation of the equation for the ordinary sibling correlation, which is the same as in Table 2.5.

Equations derived from the path diagram are given in Table 2.8. and results from the model fitting in Table 2.9.

The full model with 5 unknowns fits the data well (Table 2.9, Line 2), although not all the parameter estimates are reasonable. The simpler 3-unknown model (Line 3) fit earlier to the adoption data, a model with no genetic age change and no direct effect of parent phenotype on child environment ($t = 1.0, f = 0$), yields a good—and

TABLE 2.8 Equations for Twin-Family Studies

$$r_{T1T2} = h^2 + i^2 + c^2$$
$$r_{T1B1} = \tfrac{1}{2}\,th^2 + fc$$
$$r_{T1B2} = \tfrac{1}{2}\,th^2 + (h^2 + i^2 + c^2)fc$$
$$r_{B1B1} = \tfrac{1}{2}\,h^2 + c^2 + \tfrac{1}{2}\,cth^2(m+f)$$
$$r_{B1B2} = \tfrac{1}{4}\,h^2 + (h^2 + i^2 + c^2)f^2c^2$$

NOTE: Correlations r_{T1T2}, etc., for groups listed in Table 2.7; path symbols as in figures 2.5 and 2.6.

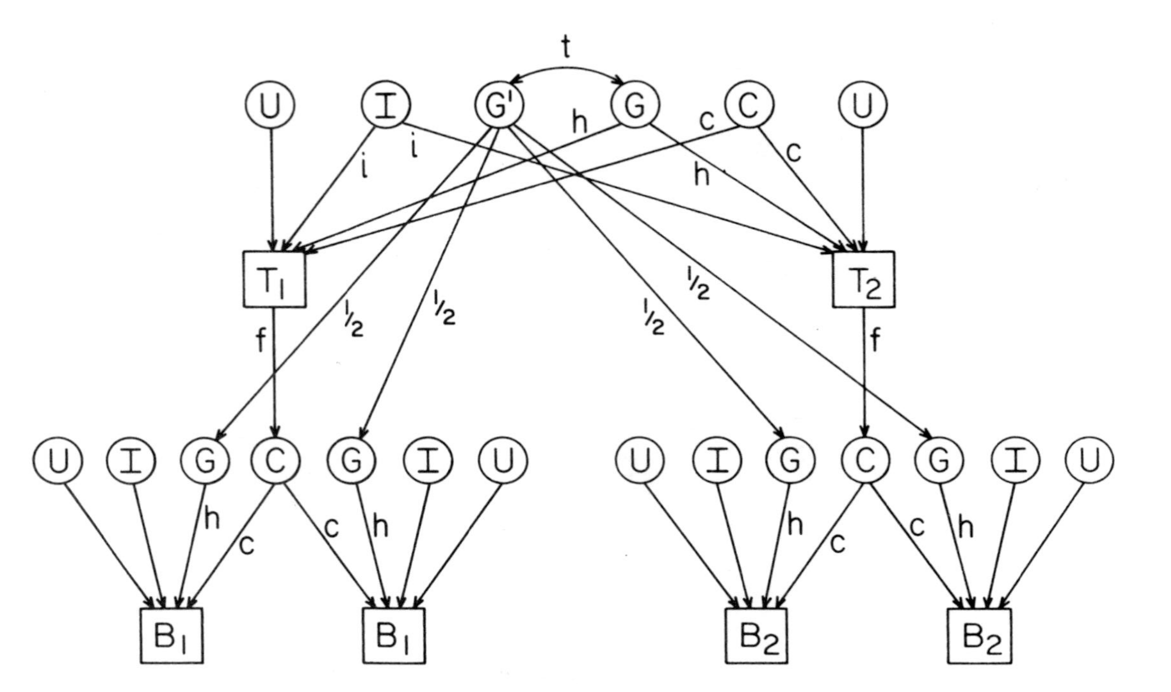

Figure 2.6. Path Model for a Twin-Family Study

T_1, T_2 = two married male MZ twins. B_1, B_2 = biological children of T_1 and T_2. Spouses of T_1 and T_2 not shown—they enter into same-family sibling correlations as in Figure 2.5 and Table 2.5. Nonadditive genetic effects assumed due to multiple-gene epistasis (I, *i*). Other symbols as in Figure 2.5. For female twins, replace *f* by *m*.

ABLE 2.9 Some Models Fit to Twin-Family Correlations for Extraversion

uestion	*Answer*	χ^2	*df*	*p*
Are *r*s from the two studies consistent?	Yes	3.35	5	> .50
Can a 5-unknown model fit across samples?	Yes[a]	3.70	5	> .50
How about just h^2, c^2, and i^2?	Yes	3.75	7	> .80
How about just h^2?	Yes	3.75	9	> .90
How about $h^2 = .366$, $c^2 = 0$, $i^2 = .142$?	Yes	6.21	10	> .70

Some paths have unreasonable values, however.

ensible—fit. Further eliminating i^2 and c^2 from this model does not ignificantly worsen its fit (Line 4). The value of h^2 for this final nodel was .44. Hewitt (1984), in a reanalysis of the complete Swedish lata set, also found that the data could be adequately fit by a one-parameter model based on h^2 alone.

Finally, we may ask whether these data are statistically compatible vith the values found to fit the twin and adoption data earlier in this hapter, namely, $h^2 = .366$, $i^2 = .142$, and $c^2 = 0$ (plus $m = f = 0$ and $t =$ .0). The answer is yes (Line 5). The χ^2_{10} from fitting these values is .21, $p > .70$; the difference from the earlier χ^2_7 is 2.46 with 3 *df*, not ignificant ($p > .30$). So far, so good—but this model will fail on the ext data set.

EPARATED TWINS

A behavior genetic design with some particular points of interest is he comparison of twins who have been reared apart in different amilies with twins who have been reared together. Because twins vho have been reared apart are rare and hard to locate, such studies re difficult to carry out. Nevertheless, there have been at least four tudies in which an Extraversion measure has been obtained for twins vho have lived for at least a substantial proportion of their childhood n separate homes. Correlations from these studies are shown in able 2.10.

One should not overstate the degree of separation of the twins in hese studies. Although many of the twins in the British and Scandinavian studies were separated within the first year of life, a few

TABLE 2.10 Extraversion Correlations in Four Studies of Separated Twins

	Finland		*Sweden*		*Minnesota*		*Britain*	
	r	*Pairs*	*r*	*Pairs*	*r*	*Pairs*	*r*	*Pai*
MZ apart	.38	30	.30	95	.34	44	.61	42
MZ together	.33	47	.54	150	.63	217	.42	43
DZ apart	.12	95	.04	220	–.07	27		
DZ together	.13	135	.06	204	.18	114		

SOURCES: Finland—Langinvainio, Kaprio, Koskenvuo, & Lönnqvist (1984); Sweden—Pedersen, Plomin, McClearn, & Friberg (1988); Minnesota—Tellegen, Lykken, Bouchard, Wilco Segal, & Rich (1988); Britain—Shields (1962).

were separated at ages as late as 10 years (Langinvainio, Kapri Koskenvuo, & Lönnqvist, 1984; Pedersen, Plomin, McClearn, & Friberg, 1988; Shields, 1962). The twins in the Minnesota study wer all separated at less than 5 years; the median age was less than months (Tellegen, Lykken, Bouchard, Wilcox, Segal, & Rich, 1988). I a number of pairs, there was some degree of contact between the twir during childhood, and in most (but not all) cases, the twins ha reunited at some point prior to their testing in adulthood. Neverthe less, at the very least, we have a group of twins who for a substantia portion of their childhood were brought up in different homes t compare to twins reared together in the same home.

First, we may ask if the data are consistent across the four sample by fitting common values of the correlations to the four data sets. Th hypothesis that the samples are homogeneous is tenable ($\chi^2_{10} = 14.2$ $p > .10$).

Path diagrams for twins reared apart are shown in Figure 2.7 an equations derived from them in Table 2.11. Model-fitting results ar shown in Table 2.12. Fitting the model with three unknown param

TABLE 2.11 Equations for Studies of Separated Twins

$$r_{MZT} = h^2 + i^2 + c^2$$
$$r_{MZA} = h^2 + i^2$$
$$r_{DZT} = \tfrac{1}{2}h^2 + c^2$$
$$r_{DZA} = \tfrac{1}{2}h^2$$

NOTE: Subscripts T and A refer to twins reared together and apart; path symbols as in Figure 2

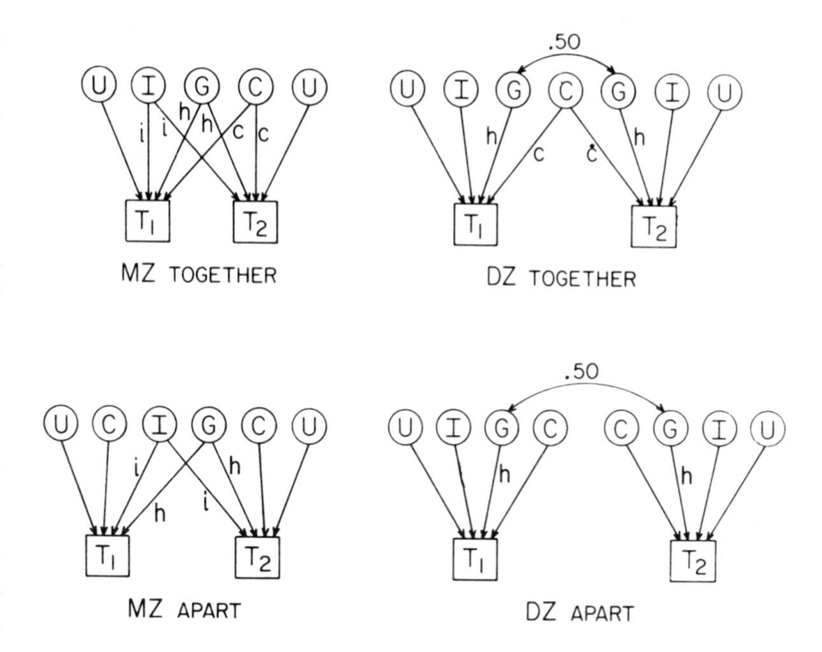

Figure 2.7. Path Models for Twins Reared Together and Apart
Nonadditive genetic effects assumed due to multiple-gene epistasis. T_1, T_2 = twins of a pair. G = additive genotype, I = epistasis, C = common environment, U = unshared environment. Paths *h*, *c*, *i* = effects of G, C, I on trait.

ters, h^2, c^2, and i^2 to the data (Table 2.12, Line 2) leads to a χ^2_{11} of 15.53, $p > .10$, a reasonable fit. The values for this solution are $h^2 = .04$, $c^2 = .12$, and $i^2 = .39$, representing a moderately large broad heritability of .04 + .39 = .43, but a very small estimate of narrow heritability. Note that c^2 is positive. It is dependably so—setting it to zero leads to a significant χ^2 (Line 3). The increase is $\chi^2_1 = 6.02$, $p < .02$. This tends to rule out certain possible models of the data for ordinary twins discussed earlier, for example, the model of parents and friends contrasting Larry and Barry, which should produce lower correlations of twins reared together than apart, rather than the higher ones observed here. It also rules out the particular set of values $h^2 = .366$, $i^2 = .142$, and $c^2 = 0$ that fit the data for ordinary twins, adoptive families, and MZ twin families.

TABLE 2.12 Some Models Fit to Extraversion Correlations from Studies of Separated Twins

Question	*Answer*	χ^2	*df*	*p*
1. Are the *r*s from the four studies consistent?	Yes	14.21	10	$> .10$
2. Will the same h^2, c^2, and i^2 fit in all four?	Yes	15.53	11	$> .10$
3. Could c^2 be set to zero?	No	21.55	12	$< .05$
4. How about h^2, i^2, $c^2_{MZ} = .12$, $c^2_S = 0$?	Yes	15.39	12	$> .20$
5. How about the preceding, but with $i^2 = 0$?	No	26.33	13	$< .02$

Are there other possibilities? Suppose that c^2 is positive, but the equal environments condition is not met, that is, that $c_{MZ} \neq c_{DZ}$. For example, let us take the values $c^2_{MZ} = .12$ and $c^2_{DZ} = .0$, which were obtained from one version considered in the regular twin design. Will a model of this sort fit the data for twins reared apart? It will, if i^2 is left free—$\chi^2_{12} = 15.39$, $p > .20$, with i^2 solved as .25. It will not, if i^2 is set to zero—$\chi^2_{13} = 26.33$, $p < .02$. Thus apparently a model with both nonadditive genetic variance and unequal twin environments may be required to fit the data.

Combined Model Fitting

Thus we are left with models with unequal twin environments and genetic nonadditivity, and we would like to see if they can fit all the data at once. Rather than proceeding in a piecemeal fashion as we have been doing, we can set up all the equations and fit them simultaneously to all the data. This will give us one overall goodness-of-fit test, one best set of estimates for the unknown parameters h^2, c^2_{MZ}, c^2_S, and i^2, and the possibility of carrying out tests of the significance of individual parameters that are based on the total data.

To illustrate this procedure, let us take the equations from tables 2.3, 2.5, 2.8, and 2.11, and the data from tables 2.1, 2.4, 2.7, and 2.10. We might reasonably include or exclude the Finnish data from the MZ-DZ comparisons; they have been excluded here in order to maintain a homogeneous data set for the twins. The Finnish data for both sexes were dropped in order to keep things more simple. Thus we are

TABLE 2.13 Some Models Fit to Extraversion Correlations from Four Twin Studies, Three Adoption Studies, Two Twin-Family Studies, and Four Studies of Separated Twins

Question	*Answer*	χ^2	*df*	*p*
1. Does model with 6 unknowns fit the data?	Yes	64.32	52	> .10
2. Could i^2 be set to zero?	Yes	.58	1	> .30
3. Equal environments? ($c^2_{MZm} = c^2_{Sm}$, $c^2_{MZf} = c^2_{Sf}$)	No	6.65	2	< .05
4. Both $i^2 = 0$ and equal environments?	No !	61.93	3	<< .001
5. Could the *c*s all be zero?	No	18.05	3	< .001
6. Could just the c_Ss be zero?	Yes	3.81	2	> .10
7. And i^2 as well?	Yes	3.81	3	> .10

NOTE: 6 unknowns in Line 1 are h^2, i^2, c^2_{MZm}, c^2_{MZf}, c^2_{Sm}, c^2_{Sf}. Remaining lines are all χ^2 differences from Line 1.

attempting simultaneously to fit data on Extraversion from four twin studies, three adoption studies, two twin-family studies, and four studies of separated twins. The results of fitting various models are summarized in Table 2.13.

The initial model we will fit makes the assumptions that we have found generally to be consistent with the Extraversion data—negligible spouse correlation, no systematic influence of parent trait on child environment ($m = f = 0$), and perfect genetic age-to-age correlation over the age ranges involved ($t = 1.0$). We will solve for h^2, i^2, c^2_{MZ}, and c^2_S, with the latter two distinct for males and females (where the sexes are pooled we use the simple average). It will be noted that in effect we are here incorporating an additional hypothesis—that the small but reliable gender differences in the twin correlations are environmental rather than genetic in origin. We will examine the merits of this assumption in due course.

The overall solution provides an acceptable fit to the combined data (Line 1 of the table). The solved-for values of h^2 and i^2 are .33 and .05, respectively, estimating a broad-sense heritability of .38 for Extraversion. The values of c^2_{MZ} are .10 for males and .15 for females; for c^2_S they are .03 and .04 respectively.

Are the assumptions of nonadditivity and of unequal twin environments both essential? The first is not. Setting i^2 to zero, we obtain a χ^2_{53} of 64.90. The increase of .58 (Line 2) over 64.32, considered as a

χ^2 with 1 *df*, is clearly nonsignificant ($p > .30$). Setting $c^2_{MZ} = c^2_S$ separately for males and females leads to a χ^2_{54} of 70.97, however. The increase of 6.65 (Line 3) over 64.32 *is* statistically significant, although not especially large. If we attempt to eliminate both nonadditivity and unequal twin environments simultaneously, we obtain a huge χ^2_{55} of 126.25, whose increase over any of the preceding χ^2s is statistically highly significant ($p << .001$). Thus the data taken as a whole do not demand both the assumption of unequal MZ and DZ environments and of nonadditive genetic variance—the former alone will do. The two assumptions tend to represent alternative hypotheses, however, and an attempt to exclude both leads to a very bad fit.

Can we get rid of the c^2s altogether? No. Setting them all to zero (Line 5) leads to a significant increase in χ^2—18.05, 3 *df*, $p < .001$—even with i^2 present. How about a common environmental factor restricted to MZ twins, with that for DZs and ordinary siblings set to zero? This will fit the data: $\chi^2_{54} = 68.13$, $p > .05$; the χ^2 increase is a nonsignificant 3.81 (Line 6). The estimates of c^2_{MZ} are .10 for males and .15 for females; h^2 is .38. Does nonadditive genetic variance make an additional contribution in this model? No. The estimate of i^2 in the preceding model is .009, and setting it to zero (Line 7) causes an entirely trivial increase in χ^2, undetectable to two decimals.

What are we to make of the fact that when we dealt just with the data for twins reared apart, a model with both unequal twin environments and nonadditive genetic variance was required in order to fit the data, whereas with the combined data set a model with one of these assumptions will suffice? Presumably, this means that a less than adequate fit in one part of the data is being offset by a more than adequate fit in other parts. Whether this is a virtue or not is somewhat a matter of scientific taste. If one believes that there may be local anomalies in particular data sets that are smoothed out when a broader range of data is considered, one will prefer the fit to the combined data. If not, one might favor a compromise between the Line 2 and Line 3 solutions, which would involve a smaller value for c^2_{MZ} and a modest i^2, but would include both.

Finally, is it necessary that the sex difference in this model be in the shared MZ environment? Alternative models with the sex difference in h^2 or i^2 were fit to the data. The results are shown in Table 2.14. In

TABLE 2.14 More Models Fit to Combined Extraversion Correlations: Sex Differences in a Model with h^2, i^2, and c^2_{MZ}

Model	χ^2	*df*	*p*
1. Sex differences on all three	67.41	52	> .05
2. Sex differences for c^2_{MZ} only	.72	2	> .50
3. Sex differences for i^2 only	1.13	2	> .50
4. Sex differences for h^2 only	.17	2	> .90
5. No sex differences	8.49	1	< .01

Difference χ^2s, except for Line 1. Differences are from Line 1, except for the last, which is with the largest of the preceding (Line 3).

the first line is shown a model with separate male and female parameters for all three of h^2, i^2, and c^2_{MZ}. Its χ^2 value of 67.41 represents a reasonable fit to the data. In each of the next three lines are shown the increase in χ^2 resulting from allowing the given parameter to differ, but requiring the other two parameters to be equal for males and females. That is, in Line 2 sex differences are allowed for c^2_{MZ} but not for i^2 and h^2, and so forth. None of these χ^2s is close to being statistically significant—that is, allowing sex differences on any one of the three parameters will do quite nicely. The last line shows that eliminating sex differences on all three parameters leads to a significant worsening of fit. Thus, the assumption that there is a sex difference in additive genetic variance, epistatic genetic variance, or special MZ environments leads to a reasonable fit to the data; one of them is necessary, but any one will do.

Conclusions

In the present chapter, we have introduced the major designs that have been applied to the study of the behavior genetics of personality traits and a general model-fitting method that can be used to summarize and analyze the correlations among genetic and social relatives that are observed. We have illustrated these methods via their application to a single personality trait, the dimension of extraversion-introversion. We have found that an assumption of some degree of nonadditivity in the genetic variance or an assumption of unequal resemblance of identical and fraternal twins' environments must be added to a simple model of genes and unshared environment in order to fit the data. The genes accounted for 35% to 39% of the individual variation in Extraversion, depending on the particular model, with shared environment accounting for 0% to 19%—values above 4% being found only for MZ twins. Remaining factors, including environmental influences not shared by family members, possible gene-environment interactions, and errors of measurement, account collectively for 46% to 63%. Because our major interest in this chapter is in illustrating the application of behavior genetic methods, we will defer further comment on the substance of these results until we can look at them in the context of other personality traits to be considered in the next chapter.

The major messages of the present chapter are, first, that a variety of different behavior genetic designs can provide evidence concerning the contribution to personality of various broad genetic and environmental classes of variation, and second, that model fitting provides a convenient and flexible way of asking questions of complex sets of data.

3

Genes, Environment, and the Big Five Personality Traits

The preceding chapter described a number of methods used in human behavior genetics, and illustrated them by application to a particular personality dimension, extraversion-introversion. In the present chapter, we will consider a broader array of personality characteristics and ask how the genes and environment contribute to their variation. Are differences among individuals in, for example, emotional stability, or kindness, or resolve more closely associated with differences in the individuals' genes or with differences in the environments to which they have been exposed?

An immediate problem is the sheer number and variety of dimensions along which individuals may vary. How many? One approach is to start with those characteristics that people have found interesting and important enough to give names to; i.e., the trait adjectives in English or another natural language (Allport, 1937).

The pioneering effort of this kind seems to have been a study done in 1933, in Germany, in which Baumgarten compiled from German dictionaries a list of 941 adjectives and 688 nouns denoting human character—a relatively modest fraction of the estimated 9,500 such terms in the German language (see John, Goldberg, & Angleitner, 1984). A short time later, Allport and Odbert (1936) went through an unabridged English dictionary and picked out 17,953 words that could be used to used to describe people, mostly adjectives. Of this total list, about one-quarter, or 4,504 words, were judged by the authors as describing "real" traits of personality, such as *aggressive* or *sociable*. The remainder included, for example, terms descriptive of temporary moods or states *(rejoicing, frantic)*, evaluations of character *(acceptable, worthy)*, explanations of behavior *(crazed)*, physical qualities *(red-headed)*, and capacities or talents *(gifted)*.

The Big Five

Of course, even 4,504 words are still far too many for practical use. There are many synonyms or near synonyms in the list, however, and many words that are rare or obsolete. A number of authors have undertaken distillations of the original list. In an early attempt of this kind, Cattell (1943, 1946) drastically pruned the list to 171 terms, which he then empirically and conceptually grouped into 35 bipolar clusters. These provided the basis for several factor analyses by various authors. A number of these analyses appeared to converge on a similar five-factor solution (e.g., Fiske, 1949; Norman, 1963). This factor solution, popularly known as The Big Five, has been the object of considerable recent research attention, and variants of it have been reported in analyses based on other reductions of Allport and Odbert's list, in the items or scales of various self-report personality inventories, in the clinical judgment items of the California Q-sort, and in different languages—German, Dutch, and Japanese. (For recent reviews of this research, see Digman, 1990; John, 1990; and McCrae, 1989.)

What are these five dimensions? Different investigations have differed slightly in the labels assigned and in the precise nature of the factors defined, but some representative labels are shown in Table 3.1. The first-listed term, in boldface, is that used by Norman, whose terminology is perhaps the most popular. The other labels have been suggested by other writers (those given in parentheses refer to the opposite end of the dimension).

TABLE 3.1 The Big Five Dimensions of Personality

I.	**Surgency,** extraversion, dominance
II.	**Agreeableness,** likability, friendliness
III.	**Conscientiousness,** conformity, will to achieve
IV.	**Emotional Stability,** (anxiety), (neuroticism)
V.	**Culture,** intellect, openness to experience

NOTE: Terms in parentheses mark opposite end of dimension.

A more concrete idea of the substance of the five dimensions may be obtained from the adjectives given in Table 3.2. These adjectives were selected by 10 judges from among the 300 items on an adjective checklist as prototypic of one or another of the Big Five factors (John, 1990). Only items which were agreed on by at least 9 of the 10 judges were included on an initial list; those that also survived a factor analysis constitute Table 3.2. (The factor analysis was based on a study in which raters applied the adjectives to a sample of 140 male and 140 female subjects.) Adjectives characterizing the opposite end of the factor from that labeled are shown in the bottom part of the table.

Examination of the list of adjectives in the table suggests that each of the Big Five factors summarizes several distinguishable aspects of behavior which tend to go together empirically. Thus the first dimension, *Surgency,* encompasses subclusters representing such traits as high activity level, self-assertion, sociality, and impulsiveness. The *Agreeableness* items, the second dimension, include some focused on emotional qualities and some on prosocial behavior. The third factor, *Conscientiousness,* contains both items emphasizing orderliness and items stressing social responsibility.

The fourth factor, *Emotional Stability,* is represented in Table 3.2 by its negative pole, emotional instability, because this is better identified by adjectives. (The stable end of the dimension was represented in the original listing by four items—*calm, contented, stable,* and *unemotional*—but they failed to survive the factor analytic screening.) The adjectives on this factor describe people in emotional difficulties, but of varying sorts, including anxiety, depression, and irritability. Simple hostility, on the other hand, tends to go on the negative pole of the Agreeableness dimension *(quarrelsome, unfriendly).* Finally, the fifth

TABLE 3.2 Adjectives Characterizing the Big Five Personality Dimensions

I. Surgency	*II. Agreeableness*	*III. Conscientiousness*	*IV. Emotional Instability*	*V. Culture*
Talkative	Sympathetic	Organized	Tense	Wide interests
Assertive	Kind	Thorough	Anxious	Imaginative
Active	Appreciative	Planful	Nervous	Intelligent
Energetic	Affectionate	Efficient	Moody	Original
Outgoing	Soft-hearted	Responsible	Worrying	Insightful
Outspoken	Warm	Reliable	Touchy	Curious
Dominant	Generous	Dependable	Fearful	Sophisticated
Forceful	Trusting	Conscientious	High-strung	Artistic
Enthusiastic	Helpful	Precise	Self-pitying	Clever
Show-off	Forgiving	Practical	Temperamental	Inventive
Sociable	Pleasant	Deliberate	Unstable	Sharp-witted
Spunky	Good-natured	Painstaking	Self-punishing	Ingenious
Adventurous	Friendly		Despondent	
Noisy	Cooperative		Emotional	
Bossy	Gentle			
	Unselfish			
	Praising			
	Sensitive			
Retiring				
Withdrawn	Hard-hearted	Slipshod		Unintelligent
Silent	Quarrelsome	Irresponsible		Shallow
Shy	Unfriendly	Frivolous		Simple
Reserved	Cold	Disorderly		Narrow interests
Quiet	Fault-finding	Careless		Commonplace

SOURCE: John (1990). Copyright Guilford Press, adapted by permission.
NOTE: The above adjectives were among 112 chosen from the 300-item Gough & Heilbrun Adjective Check List as characteristic of the dimension in question by at least 9 of a group of 10 judges. Those listed in the table survived a subsequent empirical factor analytic validation by having a loading of at least .40 on the appropriate factor and no higher on any other.

factor, *Culture,* has subsets of items focusing on intellectual facility, creativity, and intellectual motivation.

A priori, it is far from clear at what level of trait definition a behavior genetic analysis might be most fruitful. To take, for example, the Emotional Stability dimension: Do the genes set the broad outlines within which individual experiences determine which persons wind up as anxious, which irritable, and which depressed? Or do the genes act primarily to determine which of these three modes of response to adversity will characterize an individual and the general factor merely reflects how much adversity? Or would the most informative level of behavior genetic analysis be closer to that of the individual adjectives, with (for example) *tension, anxiety, nervousness,* and *fearfulness* being basically different phenomena, influenced by different (though perhaps overlapping) sets of genes? Or might there be important genetic mechanisms operating at all these levels, not to mention above or below or between them in a hierarchical sense, an arrangement that would make a complete behavior genetic analysis of the Emotional Stability dimension a very challenging task indeed?

Firm answers to the above questions do not yet exist. Nevertheless, we will proceed on the assumption that it is not necessary to know everything in order to know anything. In the present chapter, we will use the Big Five structure as a basis for dividing up the personality domain into more manageable chunks. In the next chapter, we will pursue the question of whether a more differentiated level of analysis would provide more informative behavior genetic results.

SURGENCY

We begin, then, with Factor I, Surgency.

The Big Five dimension designated as Surgency in many respects closely resembles the trait of Extraversion which we have already considered extensively in chapter 2. Indeed, some writers on the Big Five, such as McCrae and Costa (e.g., 1985, 1987), use the label extraversion to identify this dimension. Surgency, a term originally introduced by Cattell (1933), emphasizes the components of activity, energy, and enthusiasm which are part of Factor I. The term extraversion places more stress on the sociability aspect. Perhaps neither term does full justice to the social power or dominance component that is also frequently found as part of this dimension.

We will take the analysis of Extraversion as our primary information about the behavior genetics of Big Five Factor I. It will be recalled that a model solely including additive genetic variance and within-family environmental variance could be excluded, but that after that there were alternative possibilities involving nonadditive genetic variance or relaxation of the assumption of equally similar environments for MZ and DZ twins.

TABLE 3.3 Estimates from Fitting Two Simple Models to Extraversion Data

Model	h^2	i^2	c^2_{MZ}	c^2_S
1. Additive genes, MZ and sibling environments	.36		.15	.00
2. Additive genes, epistasis, equal environments	.32	.17		.02

Table 3.3 summarizes the results from two simple models, which we will find useful for comparison with other Big Five traits. One model includes additive genetic variance, special MZ twin environments, and sibling/DZ twin environments: h^2, c^2_{MZ} , and c^2_S . The other has additive and epistatic genetic variance, h^2 and i^2, plus the equal environments assumption $c^2_{MZ} = c^2_S$. The combined sexes are used. The first model gives an estimate for the additive genetic variance of .36. The other gives a slightly lower estimate of .32, but this is more than offset by an appreciable estimate of nonadditive genetic variance, .17. The c^2_{MZ} estimate (with i^2 absent) is .15, and the estimate of c^2_S is negligible for either model.

EMOTIONAL STABILITY

We consider the fourth of the Big Five factors, Emotional Stability, next, because Eysenck's Neuroticism scale, a good marker for the emotional instability end of this dimension, is available from the studies that used Eysenck's Extraversion scale, our main marker for Factor I. Thus, again, we have the luxury of a great variety of studies and of very large twin samples, plus reasonable assurance (not *no* doubts) concerning the comparability of measures across studies.

The results for twin studies, adoption studies, twin-family studies, and studies of separated twins are shown in tables 3.4 to 3.7. On the whole, there is somewhat less consistency across different studies for the Neuroticism data than there was for Extraversion. For the twins, the situation is comparable for the two traits: there is significant heterogeneity across all five studies ($\chi^2_{16} = 72.60$, $p < .001$), although not when the Finnish data are excluded ($\chi^2_{12} = 15.35$, $p > .20$). The twin-family results in Table 3.6 are also homogeneous ($\chi^2_5 = 3.96$, $p >$ .50). Unlike Extraversion, however, the adoption study results for

TABLE 3.4 Neuroticism Correlations for MZ and Same-Sex DZ Pairs in Five Large Twin Studies

	Britain		*U.S.*		*Sweden*		*Australia*		*Finland*	
	r	*Pairs*	*r*	*Pairs*	*r*	*Pairs*	*r*	*Pairs*	*r*	*Pairs*
Male twins										
MZ pairs	.51	70	.58	197	.46	2279	.46	566	.33	1027
DZ pairs	.02	47	.26	122	.21	3670	.18	351	.12	2304
Female twins										
MZ pairs	.45	233	.48	284	.54	2720	.52	1233	.43	1293
DZ pairs	.09	125	.23	190	.25	4143	.26	751	.18	2520

SOURCES: Britain—Eaves, Eysenck, & Martin (1989); U.S.—Loehlin & Nichols (1976); Sweden—Floderus-Myrhed, Pedersen, & Rasmuson (1980); Australia—Martin & Jardine (1986); Finland—Rose, Koskenvuo, Kaprio, Sarna, & Langinvainio (1988).

Neuroticism in Table 3.5 are significantly heterogeneous ($\chi^2_{12} = 24.57$, $p < .02$), as are the results from the studies of separated twins in Table 3.7 ($\chi^2_{10} = 21.25$, $p < .02$).

One can only speculate concerning the sources of this heterogeneity for the Neuroticism scales across studies. Are the differences among the different versions of the Eysenck scales greater for Neuroticism than for Extraversion? Do the determinants of emotional instability vary more from population to population than those of extraversion? Is the sampling from study to study more sensitive to the emotional

TABLE 3.5 Neuroticism Correlations in Three Adoption Studies

	Britain		*Minnesota*		*Texas*	
	r	*Pairs*	*r*	*Pairs*[a]	*r*	*Pairs*
Mother and biological child	.13	309	.21	255	.01	57
Father and biological child	.10	236	.14	255	–.13	56
Mother and adopted child	–.03	127	.12	187	–.03	257
Father and adopted child	.21	93	–.09	182	.16	247
Biologically related children	.04	418	.28	135	–.12	17
Adoptively related children	.23	58	.05	75	.09	125

SOURCES: Britain—Eaves, Eysenck, & Martin (1989); Minnesota—Scarr, Webber, Weinberg, & Wittig (1981), data from unpublished appendix, courtesy the author; Texas—Loehlin, Willerman, & Horn (1985), unpublished data.
a. Estimated.

TABLE 3.6 Neuroticism Correlations in Two Twin-Family Studies

	Sweden		*U.S.*	
	r	*Pairs*	*r*	*Pairs*
MZ twins	.45	72	.51	44
Twin and own child	.20	264[a]	.10	149
Twin and MZ twin's child	.06	71	.02	121
Siblings	.21	75[a]	.01	102
Cousins via MZ twins	.16	54	−.02	84

SOURCES: Sweden—Price, Vandenberg, Iyer, & Williams (1982); U.S.—see Loehlin (1986b).
a. Includes some in families of DZ twins.

instability of the subjects than to their extraversion? Does some other factor, such as age, matter more for Neuroticism?

Nevertheless, although the heterogeneity requires us to view the results with caution, it is still of interest to fit models to these data. Despite the less than perfect fit, comparisons of the fit of different models can still be informative, and estimates of genetic and environmental parameters may still represent useful approximations.

Table 3.8 summarizes several model-fitting runs to the same combination of samples used earlier for Extraversion: the four twin samples excluding the Finnish study, the three adoption studies, the two twin-family studies, and the four studies of separated twins. As in the case of Extraversion, omitting both the assumption of nonadditive genetic variance and the assumption of differing MZ and DZ environmental resemblance leads to a highly significant worsening of fit (Line

TABLE 3.7 Neuroticism Correlations in Four Studies of Separated Twins

	Finland		*Sweden*		*Minnesota*		*Britain*	
	r	*Pairs*	*r*	*Pairs*	*r*	*Pairs*	*r*	*Pairs*
MZ apart	.25	30	.25	95	.61	44	.53	42
MZ together	.32	47	.41	151	.54	217	.38	43
DZ apart	.11	95	.28	218	.29	27		
DZ together	.10	135	.24	204	.41	114		

SOURCES: Finland—Langinvainio, Kaprio, Koskenvuo, & Lönnqvist (1984); Sweden—Pedersen, Plomin, McClearn, & Friberg (1988); Minnesota—Tellegen, Lykken, Bouchard, Wilcox, Segal, & Rich (1988); Britain—Shields (1962).

TABLE 3.8 Some Models Fit to Neuroticism Correlations from Four Twin Studies, Three Adoption Studies, Two Twin-Family Studies, and Four Studies of Separated Twins

Question	*Answer*	χ^2	*df*	*p*
1. Does model with 6 unknowns fit the data?	No	89.73	52	< .001
2. Could i^2 be set to zero?	Yes	2.36	1	> .10
3. Equal environments? ($c^2_{MZm} = c^2_{Sm}$, $c^2_{MZf} = c^2_{Sf}$)	Yes	.50	2	> .70
4. Both $i^2 = 0$ and equal environments?	No !	35.82	3	<< .001
5. Could the *c*s all be zero?	No !	33.18	3	<< .001
6. Could just the c_Ss be zero?	No	18.74	2	< .001

NOTE: 6 unknowns in Line 1 are h^2, i^2, c^2_{MZm}, c^2_{MZf}, c^2_{Sm}, c^2_{Sf}. Remaining lines are all χ^2 differences from Line 1.

4), but only one of the two is necessary; in the case of Neuroticism either can be dispensed with (Lines 2 and 3). As with Extraversion, some degree of shared environment is required (Line 5). A difference from Extraversion is that there one could assume zero shared environments for siblings, whereas for Neuroticism one cannot—making such an assumption leads to a substantial increase in chi-square (Line 6).

For the model excluding nonadditive genetic variance ($i^2 = 0$), h^2 for Neuroticism is estimated as .31, and the c^2s are .15 and .21 for MZ males and females, and .05 and .09 for male and female DZ twins/siblings. Alternatively, with the equal environments assumption, h^2 is .30 and i^2 is .12, for a broad-sense heritability of .42, and the shared environment, c^2, is .05 for males and .10 for females.

As with Extraversion, we can ask whether it is necessary to assume an environmental source of the sex difference. Table 3.9 shows the analysis. The answer, which is not the same as for Extraversion, is that h^2 is the best bet—allowing sex differences for h^2 alone (Line 4) is nearly as good as allowing sex differences for all three parameters, but this is not true for c^2_{MZ} and only barely so for i^2 (lines 2 and 3). As the last line of the table shows, either of the latter is significantly better than not providing for sex differences at all.

Table 3.10 shows the solutions for the two simple models, for comparison with similar results for the other Big Five traits. The results suggest moderate additive genetic variance, .31 or .27, a small

TABLE 3.9 More Models Fit to Combined Neuroticism Correlations: Sex Differences in a Model with h^2, i^2, and c^2_{MZ}

Model	χ^2	*df*	*p*
1. Sex differences on all three	102.26	52	$< .001$
2. Sex differences for c^2_{MZ} only	6.25	2	$< .05$
3. Sex differences for i^2 only	5.61	2	$> .05$
4. Sex differences for h^2 only	1.56	2	$> .30$
5. No sex differences	11.11	1	$< .001$

NOTE: Difference χ^2s, except for Line 1. Differences are from Line 1, except for the last, which is with the largest of the preceding (Line 2).

effect of shared sibling environment, .05 or .07, and an intermediate estimate for the ambiguous third factor (c^2_{MZ} or i^2), of .17 or .14.

AGREEABLENESS

The remaining Big Five factors, Agreeableness, Conscientiousness, and Culture, have less data available. Furthermore, there are no single measures of these dimensions that have been widely used in behavior genetic researches, as was the case for Factors I and IV and Eysenck's Extraversion and Neuroticism scales. Costa and McCrae (1985) have published an inventory with scales explicitly intended to assess the Big Five, but as yet little behavior genetic work has been done with it. Thus, for Big Five Factor II, Agreeableness, we will make do with a scattering of studies involving various questionnaire scales thought to be related to this dimension. If the results are consistent, this is satisfactory, but if they are not, we will usually not be in a position to decide among various alternatives: Does the disagreement represent a wrong judgment concerning the classification of the scale as a

TABLE 3.10 Estimates from Fitting Two Simple Models to Neuroticism Data

Model	h^2	i^2	c^2_{MZ}	c^2_S
1. Additive genes, MZ and sibling environments	.31		.17	.05
2. Additive genes, epistasis, equal environments	.27	.14		.07

TABLE 3.11 Twin Studies of Measures Related to Agreeableness

Study	r_{MZ}	r_{DZ}
Nichols and Breland review of twin studies up to 1971 (see Osborne, 1980)		
Femininity-masculinity (7 studies)	.43	.17
McCartney, Harris, and Bernieri (1990) meta-analysis of twin studies 1967-1985		
Femininity-masculinity (9 studies)	.50	.33
Aggression (8 studies)	.49	.28
Rushton, Fulker, Neale, Nias, and Eysenck (1986) British study of adult twins		
Altruism	.53	.25
Empathy	.54	.20
Nurturance	.49	.14
Aggressiveness	.40	.04
Pedersen, Lichtenstein, Plomin, DeFaire, McClearn, and Matthews (1989) SATSA		
Hostility (twins reared together)	.33	.40
(twins reared apart)	.21	.21
Bergeman, Chipuer, Plomin, Pedersen, McClearn, Nesselroade, Costa, and McCrae (in press) SATSA		
Agreeableness (twins reared together)	.41	.23
(twins reared apart)	.15	–.03
Tellegen, Lykken, Bouchard, Wilcox, Segal, and Rich (1988) study of twins reared apart		
Aggression	.46	.06

NOTE: SATSA = Swedish Adoption/Twin Study of Aging. Number of pairs—Rushton, Fulker, Neale, Nias, & Eysenck (1986), 296 & 179; Pedersen, Lichtenstein, DeFaire, McClearn, & Matthews (1989), 120, 171 & 77, 174; Bergeman et al. (in press), approx. 99, 126 & 67, 133; Tellegen, Lykken, Bouchard, Wilcox, Segal, & Rich (1988), 44 & 27.

measure of the Agreeableness factor? Is there a difference in the heritability of the specific aspects of the individual scales (as opposed to the common factor of agreeableness)? Are there important differences among the populations studied?

Table 3.11 summarizes a number of twin studies in which measures related to Agreeableness were used. In the upper rows of the table are shown two reviews of the twin literature up to 1985, overlapping by a few years, but mostly discrete. Femininity was the only Factor II measure included in the Nichols and Breland review; the McCartney, Harris, and Bernieri (1990) review included Aggression as well, a scale identifiable with the opposite pole of the Agreeableness dimension. Inclusion of Femininity scales as markers for Agreeableness is based on the fact that the traits associated with the positive end of Factor II (*kind, affectionate, sympathetic*, etc.) tend to be stereotypically feminine, just as aggression and hostility tend to be stereotypically masculine. This factor typically shows substantial sex differences (e.g., Boyle, 1989).

In addition to the twin studies included in the two reviews, Table 3.11 shows a few more recent studies. One is a questionnaire study of British twins (Rushton, Fulker, Neale, Nias, & Eysenck, 1986). It used five questionnaire scales. The authors provide intercorrelations among the scales. Based on a factor analysis of these, four appear to load on a single factor: Altruism, Empathy, and Nurturance positively, and Aggressiveness negatively. This would appear to be an Agreeableness factor. The fifth scale in the study, which defines a factor uncorrelated with this one, was Assertiveness—presumably a Factor I scale.

The Multidimensional Personality Questionnaire used in the Minnesota study of twins reared apart (Tellegen et al., 1988) includes an Aggression scale, and correlations from that scale are also shown in Table 3.11, along with data from Hostility and Agreeableness scales in a Swedish study of twins reared together and apart.

Simple twin models of the sort discussed at the beginning of chapter 2, i.e., $h^2 = 2(r_{MZ} - r_{DZ})$, would suggest quite high heritabilities for many of these traits, but the presence of a number of r_{DZ}s that are less than half their corresponding r_{MZ}s, impossible under that model, suggests that such a simple model will not do—that, again, we will need to allow for nonadditive genetic variance or special MZ environments.

Table 3.12 presents data for Agreeableness scales from several adoption and family studies. The correlations for adoptive relatives are all small, ranging from –.03 to +.16, and centering at around +.06. The shared effects of family environment certainly do not appear to be playing a large role for this dimension, although they may be making a small contribution. The correlations for biological relatives tend to be more varied, ranging from +.02 to +.57, with a median value of +.13. If one excludes the Egyptian study as possibly reflecting major

TABLE 3.12 Adoption and Family Studies of Measures Related to Agreeableness

	Parent-Child		*Siblings*	
Study	*Biol.*	*Adop.*	*Biol.*	*Adop.*
Loehlin, Horn, and Willerman (1981) Texas adoption study				
Tender-mindedness (16PF, HSPQ, CPQ)	.06	.11		–.03
Loehlin, Willerman, and Horn (1987) follow-up of above				
Tender-mindedness (16PF)	.20	.07		.06
Loehlin, Willerman, and Horn (1985) 2nd adoption sample				
Femininity (CPI)	.03	.00		.04
Vigorous (TTS)	.33	.06		.16
Ahern, Johnson, Wilson, McClearn, and Vandenberg (1982) Hawaii family study				
Nurturance (ACL)	.09		.04	
Aggression (ACL)	.09		.02	
Femininity (CPS)	.19		.14	
Empathy (CPS)	.12		.11	
Tender-mindedness (16PF)	.20		.06	
Abdel-Rahim, Nagoshi, Johnson, and Vandenberg (1988) Egypt family study				
Femininity (CPS)	.23		.18	
Empathy (CPS)	.50		.57	

NOTE: 16PF, HSPQ, CPQ = versions of Cattell factor scales for different ages; CPI = California Psychological Inventory; TTS = Thurstone Temperament Schedule; ACL = Adjective Check List; CPS = Comrey Personality Scales. Numbers of pairings: Texas 1st 178, 409, 122; Texas follow-up 162, 358, 118; Texas 2nd, CPI 105, 494, 123, TTS 110, 528, 128; Hawaii, ACL 1710, 616, CPS 321, 100, 16PF 275, 84; Egypt 206, 47.

cultural or other population differences, the median correlation for biological relatives in Table 3.12 would be about +.10. Such a low correlation reinforces the need for caution in interpreting the twin data.

An overall joint model fitting to the data of tables 3.11 and 3.12 suggests itself. The data are not in a form to permit the full-dress methods we have used so far, but if we are willing to forego chi-squares and significance tests, we can fit some simple models of the sort we have found reasonable in other cases, and obtain estimates of the h^2s, c^2s, i^2s, etc., that they and the data imply. We will consider two

TABLE 3.13 Estimates from Fitting Two Simple Models to Agreeableness Data

Model	h^2	i^2	c^2_{MZ}	c^2_S
1. Additive genes, MZ and sibling environments	.28		.19	.09
2. Additive genes, epistasis, equal environments	.24	.11		.11

such simple models. One model includes additive genetic variance, special MZ twin environments, and sibling/DZ twin environments: h^2, c^2_{MZ}, and c^2_S. The other has additive and epistatic genetic variance, h^2 and i^2, plus the equal environments assumption $c^2_{MZ} = c^2_S$. The assumptions $m = f = 0$ and $t = 1.0$ are made throughout.

Each correlation in tables 3.11 and 3.12 was weighted equally, except that those from the reviews were weighted according to the number of studies on which they were based. This simple procedure would not be adequate for statistical testing, ignoring as it does actual sample sizes and the dependence among correlations from a given study, but it should do well enough for our purposes here. The Egyptian study was excluded. Among the unusual features of the Egyptian data were much higher levels of assortative mating than in the other populations we have considered, and our models assume this to be negligible.

The estimates from fitting the two models are shown in Table 3.13. The additive genetic variance is similar, .28 or .24, for the two models, as is the shared sibling environment, .09 or .11. The third quantity, c^2_{MZ} or i^2, depending on the model, is .19 or .11.

Which is it: Special MZ environments or nonadditive genes? Data on twins reared apart could potentially decide this issue. If such twins show DZ correlations less than half MZ correlations, this would seem to implicate nonadditive genetic factors, because special MZ environments would presumably not be operating for twins reared separately as single children. Unfortunately, the data are equivocal: the Aggression scale in the Minnesota study (Tellegen et al., 1988), shows a pattern suggestive of nonadditive genes, as does Agreeableness (weakly) in the Swedish study, but the Hostility scale in the Swedish study does not—indeed, it shows no evidence of heritability at all for twins reared together or apart. The sample sizes in the separated twin studies are small enough that the correlations are subject to a good deal of sampling fluctuation: The standard error of a correlation based on 174 pairs is about .08, and one based on 27 pairs is about .19. Changes of this magnitude in the observed correlations could have a considerable effect on interpretation.

These heritability figures, like those given elsewhere in this book, express genetic variance as a proportion of total variance, not just of reliable variance. To assess the net effects of environmental variables, one would subtract the genetic components from the test reliability, not from 1.0. Nevertheless, if these personality scales have reliabilities of .70 or so, it is clear that there remains a substantial amount of environmentally caused variance in these traits, and that the bulk of it lies within families.

CONSCIENTIOUSNESS

Again, as with Agreeableness, there is no one measure of Conscientiousness that has been widely used across a variety of different behavior genetic designs, so it is necessary to gather together information from a number of different scales that appear to lie in this general area. Tables 3.14 and 3.15 summarize results from twin studies and family and adoption studies, respectively, along the same general lines as tables 3.11 and 3.12 in the preceding section.

The first of these tables includes averages from the reviews of Nichols and Breland and of McCartney et al., (1990) along with correlations from several additional studies, including two studies of twins reared apart. Again, a number of scales show DZ correlations less than half MZ correlations, suggesting the need of nonadditive genetic or special twin parameters.

The data on adoption and family studies in Table 3.15 also bear a general resemblance to those in the corresponding table for Agreeableness. The correlations for adoptive relatives are all small, ranging from –.09 to +.13, centering on +.04 (for Agreeableness, it was +.06). Thus, again, there is not much evidence of the effects of shared family environment. Considering that Agreeableness and Conscientiousness represent areas in which fairly strong efforts at socialization occur in

TABLE 3.14 Twin Studies of Measures Related to Conscientiousness

Study	r_{MZ}	r_{DZ}
Nichols and Breland review of twin studies up to 1971 (see Osborne, 1980)		
Socialization (6 studies)	.49	.23
Conformity (5 studies)	.41	.20
McCartney, Harris, and Bernieri (1990) meta-analysis of twin studies 1967-1985		
Task orientation (7 studies)	.42	.17
Cattell, Schuerger, and Klein (1982) MAVA study, HSPQ		
Conscientiousness (G)	.50	.17
Self-control (Q3)	.56	–.07
Loehlin and Gough (1990) CPI		
Norm-favoring (vector 2)	.50	.41
Pedersen, Gatz, Plomin, Nesselroade, and McClearn (1989) SATSA		
Life direction (twins reared together)	.30	.15
(twins reared apart)	.32	.23
Bergeman, Chipuer, Plomin, Pedersen, McClearn, Nesselroade, Costa, and McCrae (in press) SATSA		
Conscientiousness (twins reared together)	.47	.11
(twins reared apart)	.19	.10
Tellegen, Lykken, Bouchard, Wilcox, Segal, and Rich (1988) study of twins reared apart		
Constraint	.57	.04

NOTE: MAVA = Multiple Abstract Variance Analysis; HSPQ = High School Personality Questionnaire; CPI = California Psychological Inventory; SATSA = Swedish Adoption/Twin Study of Aging. Number of pairs—Cattell, Schuerger, & Klein (1982), 47 and 63; Loehlin & Gough (1990), 490 & 317; Pedersen, Gatz, Plomin, Nesselroade, & McClearn (1989), 106, 124, & 66, 138; Bergeman et al. (in press), approx. 99, 126 & 67, 133; Tellegen, Lykken, Bouchard, Wilcox, Segal, & Rich (1988), 44 & 27.

families, the low average figures for the effects of shared family environment (6% and 4%) should give one pause. The correlations based on biological relationships vary more widely, from –.07 to +.45, with a median of +.14, or, excluding the Egyptian sample, +.10 (for Agreeableness, the corresponding medians were +.13 and +.10, respectively).

Simple models were fit to these data in the same manner as for Agreeableness, and the results are shown in Table 3.16. For the first model with one genetic and two environmental parameters, the Conscientiousness results of .28 and .17 resemble the .28 and .19 obtained

TABLE 3.15 Adoption and Family Studies of Measures Related to Conscientiousness

	Parent-Child		*Siblings*	
Study	*Biol.*	*Adop.*	*Biol.*	*Adop.*
Loehlin, Horn, and Willerman (1981) Texas adoption study				
Conscientiousness (16PF/HSPQ/CPQ)	.14	.10		.01
Self-control (16PF/HSPQ/CPQ)	–.05	.07		.12
Loehlin, Willerman, and Horn (1987) follow-up of above				
Conscientiousness (16PF)	–.07	.01		–.09
Self-control (16PF)	.14	.03		–.06
Loehlin, Willerman, and Horn (1985) 2nd adoption sample				
Responsibility (CPI)	.14	.05		.13
Socialization (CPI)	.11	–.02		.06
Self-control (CPI)	–.04	.05		–.03
Ahern, Johnson, Wilson, McClearn, and Vandenberg (1982) Hawaii family study				
Order (ACL)	.07		.07	
Orderliness (CPS)	.18		.19	
Social Conformity (CPS)	.23		.37	
Conscientious (G) (16PF)	.08		.17	
Controlled (Q3) (16PF)	.03		.09	
Internal Locus of Control (AQ)	.10		.12	
Cattell, Schuerger, and Klein (1982) MAVA				
Conscientiousness (HSPQ)			.20	
Self-control (HSPQ)			.10	
Abdel-Rahim, Nagoshi, Johnson, and Vandenberg (1988) Egypt family study				
Orderliness (CPS)	.26		.45	
Social conformity (CPS)	.26		.38	

NOTE: 16PF, HSPQ, CPQ = versions of Cattell factor scales for different ages; CPI = California Psychological Inventory; TTS = Thurstone Temperament Schedule; ACL = Adjective Check List; CPS = Comrey Personality Scales; AQ = Alcohol Questionnaire. Numbers of pairings: Texas 1st 178, 409, 122; Texas follow-up 162, 358, 118; Texas 2nd, CPI 105, 494, 123, TTS 110, 528, 128; Hawaii, ACL 1710, 616, CPS 321, 100, 16PF 275, 84, AQ 482,151; Cattell, 237; Egypt, 206, 47.

for Agreeableness with the same model, and the estimate of c_S^2 is a trivial .04. For the model with the two genetic and one environmental parameter the additive genetic estimate is lower, at .22, but the non-additive estimate, at .16, compensates for this, yielding approximately equal broad heritabilities of .35 and .38 for the Agreeableness and

TABLE 3.16 Estimates from Fitting Two Simple Models to Conscientiousness Data

Model	h^2	i^2	c^2_{MZ}	c^2_S
1. Additive genes, MZ and sibling environments	.28		.17	.04
2. Additive genes, epistasis, equal environments	.22	.16		.07

Conscientiousness domains. Again, the data from twins reared apart fail to suggest any compelling choice between the c^2_{MZ} or i^2 models.

CULTURE

Finally, we come to the fifth of the Big Five factors, Culture, which reflects an intellectual orientation and a receptiveness to variety of experience. As in the cases of Agreeableness and Conscientiousness, there is no single, widely used scale for this dimension, and we again consider data from a variety of studies and measures.

Table 3.17 summarizes some relevant twin data. Again, various scales appearing to lie in this general domain are included. Some intellectual and aesthetic interest scales are included from the Nichols and Breland review, along with scales assessing a personality dimension, Flexibility, which should be related to the openness aspect of the Big Five factor. The McCartney et al. (1990) review did not include a relevant category, so is not represented in the table; two twin studies using the CPI are included, however, along with the Minnesota study of twins reared apart. This comparison group of twins reared together from that study are included separately here, although they were not in the preceding tables because the reared-together groups were included in the McCartney review.

Table 3.18 summarizes family and adoption data related to Factor V, which is more scant than for the other Big Five factors. None of the adoptive correlations is high—the range is from –.02 to +.11—but the median value, +.08, is a trifle above the +.06 and +.04 in the cases of Agreeableness and Conscientiousness. As usual, the correlations between biological relatives vary over a considerable range, from +.01 to +.29; the median, +.09, is about the same here as for the others.

TABLE 3.17 Twin Studies of Measures Related to Culture/Openness

Study	r_{MZ}	r_{DZ}
Nichols and Breland review of twin studies up to 1971 (see Osborne, 1980)		
Flexibility (7 studies)	.46	.27
Artistic interests (16 studies)	.50	.32
Science interests (15 studies)	.54	.29
Loehlin and Nichols (1976) National Merit twins		
Achievement via independence (CPI)	.54	.41
Flexibility (CPI)	.48	.21
Horn, Plomin, and Rosenman (1976) Veteran twins		
Achievement via independence (CPI)	.49	.25
Flexibility (CPI)	.49	.10
Bergeman, Chipuer, Plomin, Pedersen, McClearn, Nesselroade, Costa, and McCrae (in press) SATSA		
Openness (twins reared together)	.51	.14
(twins reared apart)	.43	.23
Tellegen, Lykken, Bouchard, Wilcox, Segal, and Rich (1988) twins apart and together		
Absorption (MPQ) (twins reared together)	.49	.41
(twins reared apart)	.61	.21

NOTE: CPI = California Psychological Inventory; MPQ = Multidimensional Personality Questionnaire. Number of pairs—Loehlin & Nichols (1976), 490 & 317; Horn, Plomin, & Rosenman (1976), 99 & 99; Bergeman et al. (in press), approx. 99, 126 & 67, 133; Tellegen, Lykken, Bouchard, Wilcox, Segal, & Rich (1988), 217 & 114 and 44 & 27.

Table 3.19 shows the estimates from fitting the same two simple models as before. In either case, the result is a fairly substantial h^2, estimated at .46 or .43, a small c_S^2 of .05 or .06, and an equally small or smaller c_{MZ}^2 or i^2.

This analysis gives a heavy weight to the twin studies of interests represented in the Nichols and Breland review, because of their relatively large number (16 and 15 for artistic and science interests respectively). Do some of the distinctive features of its results depend on this? As a check, the analysis of the first model was rerun, weighting these two groups of studies as if they consisted of only three studies each. The results were little changed (.41, .03, .05 instead of .46, .05, .05), so we may conclude that they are not simply an artifact of the inclusion of a large number of studies of interests in the original analysis.

TABLE 3.18 Adoption and Family Studies of Measures Related to Culture/Openness

	Parent-Child		*Siblings*	
Study	*Biol.*	*Adop.*	*Biol.*	*Adop.*
Loehlin, Willerman, and Horn (1987) Texas adoption study follow-up				
Imaginative (16PF)	.24	.05		.08
Experimenting (16PF)	.04	.08		.06
Loehlin, Willerman, and Horn (1985) 2nd adoption sample				
Achievement via independence (CPI)	.25	.08		–.02
Flexibility (CPI)	.20	.11		.10
Ahern, Johnson, Wilson, McClearn, and Vandenberg (1982) Hawaii family study				
Intraception (ACL)	.09		.02	
Imaginative (16PF)	.29		.01	
Experimenting (16PF)	.14		.09	

NOTE: 16PF = Cattell Sixteen Personality Factor Scales; CPI = California Psychological Inventory; ACL = Adjective Check List. Numbers of pairings: Texas follow-up 162, 358, 118; Texas 2nd 105, 494, 123; Hawaii, ACL 1710, 616, CPS 321, 100, 16PF 275, 84.

The Big Five Compared

Table 3.20 gathers together the estimates from fitting the two simple models to each of the Big Five factors. Several points stand out in this table. First, the additive genetic component is always the largest, usually followed by the i^2 or c^2_{MZ} component, with c^2_S third. Culture/Openness is the most deviant of the group, with the largest estimates for additive genetic variance and the smallest estimate for c^2_{MZ} or i^2 of any of the Big Five. For the first four factors, adding a second genetic parameter in the second model leads to a decrease in

TABLE 3.19 Estimates from Fitting Two Simple Models to Culture/Openness Data

Model	h^2	i^2	c^2_{MZ}	c^2_S
1. Additive genes, MZ and sibling environments	.46		.05	.05
2. Additive genes, epistasis, equal environments	.43	.02		.06

TABLE 3.20 Summary: Estimates from Fitting Two Simple Models to Data from Big Five Traits

Big Five Factor	h^2	c^2_{MZ}	c^2_S	h^2	i^2	c^2_S
I. Surgency	.36	.15	.00	.32	.17	.02
II. Agreeableness	.28	.19	.09	.24	.11	.11
III. Conscientiousness	.28	.17	.04	.22	.16	.07
IV. Emotional Stability	.31	.17	.05	.27	.14	.07
V. Culture/Openness	.46	.05	.05	.43	.02	.06

the narrow heritability and an increase in the broad heritability over the first model. The equated MZ/DZ/sib parameter in the second model tends to lie between the two separate estimates for these two parameters in the first.

All told, then, we see appreciable effects of additive genes in the 22% to 46% range across the Big Five factors, highest for Culture, next for Surgency, and lower for the other three. We see positive, but small effects of shared sibling environment alone—negligible for Surgency; 4% to 11% for the rest. And we have an appreciable, but ambiguous, third factor, which might represent either nonadditive effects of the genes or the effect of special MZ environmental resemblance; it accounts for 11% to 19% of the variance on the first four of the Big Five factors and a lesser 2% to 5% on the last. Finally, we must not forget the variance not so far accounted for—the 44% to 55% attributable to individual environments, gene-environment interaction, and errors of measurement.

All of these estimates must be taken as tentative. Those for Surgency and Emotional Stability are based on quite extensive data, but depend on the assumption that Eysenck's Extraversion and Neuroticism scales are adequate measures of these factors. Those for the other three factors are based on somewhat miscellaneous collections of scales and rough model fitting. There may be important differences across populations (as suggested by the unusually high familial correlations in the Egyptian sample). There are differences between the sexes for Extraversion and Neuroticism, in the direction of slightly higher shared effects for females—which could be genetic or environmental. The data were not adequate for the exploration of sex differences with respect to the other three Big Five factors, but it would not, of course, be surprising if some such differences exist.

Conclusions

In this chapter, we looked at heredity-environment analyses at the level of the Big Five trait domains. For two, Surgency and Emotional Stability, we used the extensive data available from Eysenck's Extraversion and Neuroticism scales. For the other three, Agreeableness, Conscientiousness, and Culture, we summarized data from a variety of scales used in twin, adoption, and family studies.

In general, we found that the data could reasonably be described by either of two models, one involving additive genes, genetic epistasis, and shared twin/sibling environments; the other involving additive genes, special MZ twin environments, and shared environments of DZ twins and other siblings. Additive genetic effects under either model fell generally in the moderate range, 22% to 46%; shared family environmental effects were small, 0% to 11%; and the ambiguous third factor—nonadditive genetic or special MZ twin environments—was intermediate at 11% to 19%, except for Culture, where it was a lesser 2% to 5%. The remaining variation, 44% to 55%, presumably represents some combination of environmental effects unique to the individual, genotype-environment interaction, and measurement error.

Levels of Analysis

General or Specific?

In the preceding chapter, we considered the heritability of traits at the broad level of the Big Five factors. This provided a convenient summary, but this is obviously not the only possible level for behavior genetic analysis. We will illustrate this point using Big Five Factor I, the broad Surgency/Extraversion/Dominance dimension, and some of its subcomponents.

An early study of this kind involved an analysis of an Extraversion scale and two subscales representing Sociability and Impulsivity (Eaves & Eysenck, 1975). The analysis was carried out on the responses of 837 pairs of adult twins to an 80-item personality questionnaire, of which 13 items were identified as providing a Sociability scale, and 9 an Impulsiveness scale. Heredity-environment models were fit to the two scales, their sum, which can be considered an overall measure of extraversion, and their difference, a measure of the contrast between the two traits.

Very briefly, a simple model of additive genes and within-family environment, the same for the two sexes, fit the data for Sociability and Impulsiveness, as well as their sum and their difference. The estimated value of h^2 was .46 for Sociability, .36 for Impulsiveness, and .42 for their sum (Extraversion). Also of interest, the difference between Impulsiveness and Sociability was itself heritable (h^2 = .34), suggesting that the distinctive aspects of these traits, as well as what they share, are influenced by genes.

Three Aspects of Extraversion

IMPULSIVITY, DOMINANCE, SOCIABILITY

A few years later, Carey and Rice (1983) carried out a model-fitting analysis of three traits in the Extraversion domain: Social Potency or Dominance, Social Closeness, and Impulsivity versus control. (The last could be considered also to share variance with Conscientiousness, another of the Big Five, but we will follow Carey and Rice in considering it an aspect of Extraversion.) The three traits in question had been measured in various relevant groups: the adoption study of Scarr et al. (1981), twin data reported by Lykken, Tellegen, and DeRubeis (1978), and unpublished family data gathered by Carey. The three scales were from Tellegen's Differential Personality Questionnaire (DPQ). Rather than describe Carey and Rice's results in detail, I will report a combined analysis of their data with data for three similar scales from the Thurstone Temperament Schedule (TTS), labeled Dominant, Sociable, and Impulsive, which I gathered together for a model-fitting analysis (Loehlin, 1986b) from two different twin studies (Rosenman, Rahe, Borhani, and Feinleib, 1976; Vandenberg, 1962), an adoption study (Loehlin et al, 1985), and an unpublished twin-family study. To these two collections of correlations are added correlations of MZ and DZ twins reared apart from the study of Tellegen et al. (1988), using the current version of the DPQ, the Multidimensional Personality Questionnaire.

The various groups are summarized in Table 4.1, and the actual correlations are shown in Table 4.2. Table 4.1 also gives the equations used for the different groups in the model fitting. These equations include three genetic parameters: h^2, representing the additive effect of genes, d^2, representing the effects of genetic dominance, and i^2, for epistatic effects based on multiple-gene configurations. Three parameters representing shared environment are included: c^2_{MZ} , the environment shared by identical twins, c^2_S , the environment shared by other

TABLE 4.1 Multigroup Analysis of Impulsivity, Dominance, and Sociability Scales

Group	*Samples*	*Pairs*	*Equation*
MZ twins reared together	5	422	$h^2 + d^2 + i^2 + c^2_{MZ}$
MZ twins reared apart	1	44	$h^2 + d^2 + i^2$
DZ twins reared together	4	259	$\frac{1}{2} h^2 + \frac{1}{4} d^2 + c^2_S$
DZ twins reared apart	1	27	$\frac{1}{2} h^2 + \frac{1}{4} d^2$
Siblings	4	525	$\frac{1}{2} h^2 + \frac{1}{4} d^2 + c^2_S$
Offspring of MZ twins	1	84	$\frac{1}{4} h^2$
Adoptive siblings	2	218	c^2_S
Parent and biological child	6	1014	$\frac{1}{2} h^2 + c^2_P$
MZ and twin's child	1	121	$\frac{1}{2} h^2$
Parent and adopted child	5	907	c^2_P

NOTE: Some *N*s estimated, some averages. h^2 = additive genes, i^2 = multiple-gene epistasis, d^2 = genetic dominance, c^2_{MZ} = shared environment of MZ twins, c^2_S = shared environment of DZ twins or siblings, c^2_P = shared environment, parent and child.

siblings, and c^2_P, the environment shared by parent and child. The last is a simplification, in which c_P serves in lieu of the paths via the *m* and *f* parameters of the models in the preceding chapters. The distinction is not of practical import, for this parameter will in due course be set to zero as *m* and *f* were earlier. The temporal genetic correlation *t* is assumed to be 1.0, and thus is omitted from the equations. The equations assume that spouses are negligibly correlated on these three traits, an assumption supported by the marital correlations of –.03, –.03, and .12 reported by Carey and Rice (1983). The equations also assume that parameters can be equated across groups—for example, for males and females, and for persons of different ages (all of these samples consist of high school age and older, so this is not a matter of equating adults and young children). The degree of shared environment is allowed to differ for MZ twins, other siblings, and parents and offspring; it is assumed to be the same in adoptive and ordinary families, however. Genetic parameters are assumed to be the same across all types of relationships.

Most likely, none of these assumptions is exactly true. For example, in chapter 2, statistically significant sex differences were shown in the analysis of Extraversion, but they were quantitatively small and required very large samples for their dependable detection. No doubt

TABLE 4.2 Correlations for Impulsivity, Dominance, and Sociability Scales, Plus Mean Correlation in 30 Groups

Group	*IMP*	*DOM*	*SOC*	*Mean*	*Pairs*
Michigan MZ twins	.39	.56	.47	.48	45
Veterans MZ twins	.52	.58	.45	.52	102
Twin-family MZ twins	.35	.37	.30	.34	44
Minnesota MZm twins	.52	.46	.45	.48	79
Minnesota MZf twins	.37	.69	.49	.53	152
MZ twins reared apart	.50	.56	.29	.46	44
Michigan DZ twins	–.11	.27	.00	.06	34
Veterans DZ twins	.02	.03	.08	.04	119
Minnesota DZm twins	–.24	.47	.05	.10	34
Minnesota DZf twins	.15	.00	.54	.25	72
DZ twins reared apart	.03	.27	.30	.20	27
Twin-family sibs	.07	.21	.17	.15	102
Minnesota mm sibs	.04	.21	.21	.15	92[a,b]
Minnesota mf sibs	.17	.17	–.01	.11	182[a,c]
Minnesota ff sibs	.16	.13	.21	.17	149[a,d]
Twin-family, cousins via MZs	.06	.35	.12	.18	84
Texas adoptive sibs	–.02	–.02	–.07	–.04	143
Minnesota adoptive sibs	.05	.07	.13	.08	75[a]
Texas parent-child	.10	.11	.17	.13	118
Twin-family, MZ-own child	.09	.22	.22	.18	149
Minnesota father-son	.14	.18	.11	.14	142[a,e]
Minnesota father-daughter	.13	.29	.17	.20	201[a,f]
Minnesota mother-son	.05	.14	.26	.15	178[a,g]
Minnesota mother-daughter	.15	.12	.17	.15	227[a,h]
Twin-family, MZ-twin's child	.18	.24	.30	.24	121
Texas, parent-adopted child	.04	.01	.02	.02	541
Minnesota, father-adopted son	.06	.26	–.03	.10	85[a]
Minnesota, father-adopted daughter	–.12	.06	.04	–.01	98[a]
Minnesota, mother-adopted son	–.11	.04	.03	–.01	85[a]
Minnesota, mother-adopted daughter	.05	.00	.12	.06	98[a]

NOTE: IMP = impulsivity, DOM = dominance, SOC = sociability. m, f = male, female. Data from Loehlin (1986b): Michigan = Michigan high school twins (Vandenberg, 1962), Veterans = twins among U.S. veterans of World War II (Rahe, Hervig, & Rosenman, 1978), Texas = Texas adoptive families (Loehlin, Willerman, & Horn, 1985), Twin-family = unpublished twin-family study based on veterans twins. Minnesota = data from Carey & Rice (1983) and Scarr, Webber, Weinberg, & Wittig (1981): twins (Lykken, Tellegen, & DeRubeis, 1978), adoptees (Scarr et al., 1981), siblings and parent-child pooled from Carey, unpublished (see Carey & Rice, 1983), and Scarr et al. (1981). Twins reared apart: data from Tellegen et al. (1988).

a. Estimated. The following are the *N*s for the three scales, when not the same (harmonic means are in the table): b. 104, 94, 80; c. 164, 183, 202; d. 154, 158, 136; e. 131, 173, 129; f. 215, 233, 167; g. 176, 195, 166; h. 271, 206, 215.

TABLE 4.3 Models Fit to Combined Data on Impulsivity

Model parameters	χ^2	*df*	*p*	χ^2_{diff}	*df*	*p*
1. $h^2\ d^2\ i^2\ c^2_{MZ}\ c^2_S\ c^2_P$	17.09	24	> .80			
2. $h^2\ i^2$	18.04	28	> .90	.95	4	> .90
3. $h^2\ c^2_{MZ}$	21.99	28	> .70	4.90	4	> .20
4. h^2	33.10	29	> .20	11.11	1	< .01

NOTE: Comparisons in χ^2_{diff} column are with the 6-parameter model in Line 1, except the last, which is with the worse fitting of the 2-parameter models.

age can make a difference, too, yet if one looks at the twin samples in Table 4.2, there do not appear to be any consistent and dramatic differences in correlation between the middle-aged twins in the veterans and twin-family samples, the college-age twins in the Minnesota samples, and the high school twins in Michigan. Sex differences in the Minnesota samples, particularly the DZs, appear more striking (and were found to be statistically significant by Carey and Rice at several points in their model fitting). Yet these correlations are based on small samples, and the differences could well be exaggerated by selection or sampling fluctuation. The pattern of correlations for the Minnesota DZ males is not reflected, for example, in the veterans sample, which is also male. Probably the most parsimonious interpretation is sampling fluctuation, but in any case our model fitting will provide a test: If models that assume no sex differences in parameters cannot fit the data, they will be rejected.

Tables 4.3 to 4.6 show the results of various model-fitting runs, four for each table. In the first line is shown a full model that solves for all of the three genetic and the three environmental parameters that appear in the equations of Table 4.1. In the second and third lines, the additive genetic parameter h^2 is paired with either a nonadditive genetic parameter (the epistatic parameter i^2) or a parameter allowing for a special environmental resemblance for MZ twins (c^2_{MZ}). And finally, in the last line of the table, h^2 stands alone.

The information for each run may best be understood by looking at a particular table—let us take Impulsivity, Table 4.3. The χ^2 for the goodness of fit of each model is given in the first column of the table, and the associated degrees of freedom and probability values are

TABLE 4.4 Models Fit to Combined Data on Dominance

Model parameters	χ^2	*df*	*p*	χ^2_{diff}	*df*	*p*
1. $h^2\, d^2\, i^2\, c^2_{MZ}\, c^2_S\, c^2_P$	36.85	24	> .05			
2. $h^2\, i^2$	38.15	28	> .05	1.30	4	> .80
3. $h^2\, c^2_{MZ}$	40.45	28	> .05	3.60	4	> .30
4. h^2	54.34	29	< .01	13.89	1	< .01

NOTE: Comparisons in χ^2_{diff} column are with the 6-parameter model in Line 1, except the last, which is with the worse fitting of the 2-parameter models.

given in columns 2 and 3. The χ^2 of 17.09 in Line 1 of the table is based on 24 degrees of freedom—the 30 correlations that are being fit, less the 6 unknowns that are being solved for. The probability that a χ^2 this large would result from random sampling fluctuation is over .80, meaning that the fit of the model to the data may be considered to be excellent. There might be more economical models that could do just as well, however; the remaining lines in the table explore this possibility. In lines 2 and 3, we learn that two two-parameter models fit the data well in absolute terms ($p > .90$ and $p > .70$), and not significantly worse than the full six-parameter model. In the last line of the table, we find that a one-parameter model featuring h^2 alone does not fit the data too badly ($p > .20$), but it is clearly worse than even the weaker of the two-parameter models, by a χ^2_{diff} of 11.11 for 1 *df*, $p < .01$.

Looking at all three tables, 4.3 to 4.5, for Impulsivity, Dominance, and Sociability, we see that although the general levels of fit vary (they are best for Impulsivity, worst, but still acceptable, for Dominance), many features of the solutions are similar. For all three traits, the two two-parameter models are not significantly inferior to the original six-parameter model. Where the three differ is in the one-parameter model involving only h^2, which fits significantly worse than the two-parameter models for Impulsivity and Dominance, but not for Sociability.

In short, a simple additive genetic model provides a reasonable account of resemblance across these various groups in the case of Sociability, but for Impulsivity or Dominance, the assumption of nonadditive genetic variation or special environmental similarity for MZ twins significantly improves the fit to the data. On the whole,

TABLE 4.5 Models Fit to Combined Data on Sociability

Model parameters	χ^2	*df*	*p*	χ^2_{diff}	*df*	*p*
1. $h^2\,d^2\,i^2\,c^2_{MZ}\,c^2_S\,c^2_P$	31.30	24	> .10			
2. $h^2\,i^2$	33.94	28	> .20	2.64	4	> .50
3. $h^2\,c^2_{MZ}$	32.82	28	> .20	1.52	4	> .80
4. h^2	35.71	29	> .10	2.89	1	> .05

NOTE: Comparisons in χ^2_{diff} column are with the 6-parameter model in Line 1, except the last, which is with the best fitting of the 2-parameter models.

epistasis does a little better than a special MZ environment, but either is statistically acceptable.

THE THREE TRAITS COMBINED

What happens when we average correlations across these three traits to obtain a rough approximation to an overall Extraversion measure? The results are shown in Table 4.6. In the first place, the fits to the data become extremely good ($p > .99$ in several instances), probably due to the averaging out of sampling irregularities. Otherwise, the results generally resemble those in the preceding tables: The one-parameter model is significantly weaker, and the two-parameter models h^2i^2 and $h^2c^2_{MZ}$ do nearly as well as the six-parameter full model.

Carey and Rice carried out a similar averaged analysis in their report on part of these data (1983), which they mention in a footnote. They concluded that a model with just h^2 would fit the composite Extraversion trait quite well ($p > .50$), but they did not, apparently, ask whether adding a second parameter would make it fit even better.

HOMOGENEITY ACROSS SCALES AND STUDIES

Are the preceding results homogeneous across the two sets of scales and studies? That is, are the Impulsivity versus Control, Social Potency, and Social Closeness scales of the DPQ really equivalent to the Impulsive, Dominant, and Sociable scales of the TTS, and are the samples in the studies using the DPQ equivalent to those in the studies

TABLE 4.6 Models Fit to Combined Data on Extraversion (Average of Three Traits)

Model parameters	χ^2	*df*	*p*	χ^2_{diff}	*df*	*p*
1. $h^2\, d^2\, i^2\, c^2_{MZ}\, c^2_S\, c^2_P$	8.72	24	> .99			
2. $h^2\, i^2$	9.66	28	> .99	.94	4	> .90
3. $h^2\, c^2_{MZ}$	10.69	28	> .99	1.97	4	> .70
4. h^2	19.23	29	> .90	8.54	1	< .01

NOTE: Comparisons in χ^2_{diff} column are with the 6-parameter model in Line 1, except the last, which is with the worse fitting of the 2-parameter models.

using the TTS? We can test the equivalence by comparing the fits of models that estimate separate parameters in the DPQ and the TTS portions of the combined data set. If either scales or samples are *not* equivalent, we should obtain a significant improvement in χ^2 by fitting parameters separately in the two subportions of the data.

Table 4.7 shows the results. For each of the four models, two sets of χ^2s are shown: one for fitting the same parameters for both questionnaires, and one for fitting different parameters for each. As the associated probability values show, in no case do the differences in χ^2 even

TABLE 4.7 χ^2s Based on Same Versus Different Parameters for the Two Questionnaires

Model	*S/D*	*df*	*IMP*	*DOM*	*SOC*	*EXT*
1. $h^2\, d^2\, i^2\, c^2_{MZ}\, c^2_S\, c^2_P$	S	24	17.09	36.85	31.30	8.72
	D	18	14.64	32.02	27.86	6.63
			>.70	>.50	>.70	>.90
2. $h^2\, i^2$	S	28	18.04	38.15	33.94	9.66
	D	26	17.51	36.55	33.80	9.44
			>.70	>.30	>.90	>.70
3. $h^2\, c^2_{MZ}$	S	28	21.99	40.45	32.82	10.69
	D	26	20.68	38.57	32.30	10.28
			>.50	>.30	>.70	>.70
4. h^2	S	29	33.10	54.34	35.71	19.23
	D	28	33.10	54.04	35.70	19.19
			>.95	>.50	>.90	>.70

NOTE: S = same, D = different. Probability that this large a difference between S and D would occur by chance shown below each pair of S and D values. IMP = Impulsivity, DOM = Dominance, SOC = Sociability, EXT = Extraversion (averaged correlations).

TABLE 4.8 Parameter Estimates for Some Two- and One-Parameter Solutions for Extraversion-Related Traits

Trait	h^2	i^2	h^2	c^2_{MZ}	h^2
Impulsivity	.21	.23	.24	.20	—
Dominance	.36	.22	.38	.20	—
Sociability	.36	.08	.35	.10	.40
Extraversion	.31	.18	.33	.17	—

NOTE: Extraversion parameters are from fitting to the mean of correlations for the other three. Absence of data means that model fit was significantly worse than others shown.

approach statistical significance. Thus we have an assurance that the combined analysis is an appropriate one.

Finally, in Table 4.8 are presented the actual parameter values for the more promising fits to the combined DPQ/TTS data. Generally speaking, the narrow-sense heritabilities of these traits are estimated at what we might describe as a modest but respectable 20 to 40%. Impulsivity seems to be a bit lower than the others. Another 20% of the variation for Impulsivity or Dominance might be either genetic but nonadditive, or due to a special environmental resemblance applying to MZ twins alone. This component apparently plays a lesser role for Sociability. The remaining variation of these traits, from 40% to 60%, reflects measurement error, unshared environmental effects, and genotype-environment interaction. Measurement error may account for about half of this; estimates of error variance for the three Thurstone scales were obtained in one of the subsamples, and they were 30%, 12%, and 24% for Impulsivity, Dominance, and Sociability (Loehlin, 1986b). The distribution of the remainder of the variance between genotype-environment interaction and individual experience is not known, although it seems likely that individual environmental effects are substantial for personality (Plomin & Daniels, 1987).

How do these results speak to the question of the optimum level for the behavior genetic analysis of personality? Insofar as there appear to be differences among the three subtraits that are homogenized when they are combined, the results would argue for the analysis being extended to a more specific level than that of the Big Five. (They do not, of course, rule out performing analyses at both levels.) But should we stop at Impulsivity, Dominance, and Sociability, or move downward to a still finer-grained level of analysis, for example, to narrow item clusters or even single items?

The Heritability of Questionnaire Items

Are personality questionnaire *items* differently heritable? The question remains open—the topic has been addressed, but not yet fully explored. On the negative side, there have been at least two failures to replicate twin-study heritability estimates for small item clusters and single questionnaire items (Loehlin, 1986a, b). In the first of these, item clusters were originally derived from the Thurstone Temperament Schedule and, using data from the Michigan sample of high-school-aged twins, were classified as being of higher or lower heritability (Loehlin, 1965). The two groups of clusters were later compared (Loehlin, 1986b) using data from the same test given to the U.S. veterans sample of middle-aged male MZ and DZ twins. Of seven clusters originally identified as high in heritability, four now came out high and three low. Of the seven originally identified as low, the figures were three and four. Obviously, the consistency across samples did not exceed chance. The second study took individual items from the California Psychological Inventory that had been classified in the veterans twin sample as being of relatively high or low heritability (Horn, Plomin, & Rosenman, 1976), and examined them in the National Merit twin sample, using the same criteria of classification (Loehlin, 1986a). Again, there was no significant consistency.

Now, obviously, there were age and other differences between the samples in each case, which might have contributed to the inconsistency, but the theoretical possibility that behavior genetic analysis might be most informative at a highly specific level gets little encouragement from these particular analyses.

Neale, Rushton, and Fulker (1986) provide an item-level gene-environment analysis of Eysenck's EPQ that shows more promise.

The study was based on 627 pairs of adult British twins. The model-fitting program estimated h^2 and c^2 for each item and provided standard errors for these estimates. Some of the differences among h^2 estimates were large compared to the standard errors, suggesting that the results ought to be replicable. The c^2 estimates tended to be small, with relatively few as large as their standard errors, suggesting less likelihood of replication of differences here.

And do the h^2 differences replicate? Heath, Jardine, Eaves, and Martin (1989) subsequently carried out another item-level analysis of the EPQ, based on the Australian twin sample of 3,810 pairs. They did not do quite the same analysis. They estimated a dominance parameter d^2 in addition to h^2 and c^2, separately for males and females. Nevertheless, since $h^2 + 1.5d^2$ in their study should correspond to h^2 in the Neale et al. (1986) study (each would be expected to equal twice the difference between r_{MZ} and r_{DZ}), we are in position to make at least an initial check on the Neale et al. (1986) results. We will do this first for the 21 Extraversion items from the EPQ, as a logical downward extension of our analyses for this trait at the Big Five and subscale levels. For each item, values of $h^2 + 1.5d^2$ were calculated separately for males and females from the Heath et al. (1989) estimates and averaged. These values were then correlated across items with the corresponding h^2 estimates from the British study. This correlation proved to be +.35. The correlation between the male and female subsamples in the Australian data was +.36. Neither of these correlations suggests a high degree of replicability for heritability estimates of items, but they do suggest some.

Unfortunately, a similar analysis of the EPQ Neuroticism scale items provides less encouragement. The correlation between the British and Australian estimates across the 23 Neuroticism items remains positive, but drops to +.12, and that between males and females in the Australian sample goes down to +.07. This does not represent enough consistency to make such an analysis very meaningful, except perhaps with immensely large samples.

Corresponding item-level studies for Neuroticism and Extraversion in the Swedish and Finnish twin samples have not been reported, so far as I am aware, but the large samples used in these studies might lead to a clearer picture of the potential of this approach.

Conclusions

In this chapter, we considered analyses of three specific traits within the Extraversion domain, Impulsivity, Dominance, and Sociability. We found some differences among these, and cross-study consistency, suggesting that continued analysis below the level of the Big Five may be profitable. Finally, we descended to the level of individual questionnaire items and item clusters, finding a little evidence of consistency, but far from an assured replicability of results.

5

Heredity, Environment, and Change

In this chapter, we will focus on time and change. Are personality traits more heritable in infancy, childhood, adolescence, adulthood, or old age? Do individual family members become more like each other during development or more different? In either case, what can we say about the relative contribution of the genes and the environment to this process?

A key fact: Changes in trait scores can be subjected to the same sorts of heredity-environment analysis as trait scores themselves. If maturational change in a trait follows an underlying genetic schedule, and if this schedule differs across individuals, then MZ twins' changes should be more alike than DZ twins' changes, and the changes of siblings more alike than those of unrelated adopted children. If, on the contrary, trait changes in individuals are primarily due to changes in the shared family environment (such as, for example, a developing discord between the parents), then children living in the same family, whether biologically related to one another or not, should show similar changes, and these should be different from those of children in other families. Any of the methods described in chapter 2 can in principle be applied to assessing genetic or environmental influences on trait changes, merely by applying them to change scores rather than to single-occasion scores. The phrase "in principle" is important;

for a number of reasons, in practice it may be considerably more difficult to analyze changes.

First, repeated data on the same set of subjects are much harder to obtain than single-occasion data; subjects are missing on one or more occasions of testing or drop out altogether.

Second, if the change is assessed over a substantial period of time, especially if this is early in life, thorny questions arise concerning the comparability of the measurements at the different times. Sociability in a 2-year-old is manifested by a different set of behaviors than sociability in a 16-year-old, even though there may be enough continuity between the two to justify the assumption that they are in some sense "the same thing." Also the measurement options differ—the 16-year-old can fill out a printed questionnaire or respond to an interview, a 2-year-old cannot. A parent's rating of behavior can be made for either, but the parent is likely to sample the individual's social behavior quite differently in the two cases.

Third, psychometric and statistical problems bedevil change scores—most notably, the unreliability due to errors of measurement entering at each end, rather than just once as with ordinary scores. Consequently, the samples needed for comparable precision of analysis are larger than those that, due to the difficulty and expense of gathering longitudinal data, tend to be available.

For these reasons, only a modest amount of data is available for the direct behavior genetic analysis of personality change. But there is some, and we will discuss it. There is considerably more evidence of a cross-sectional kind, in the form of heritability analyses carried out at different ages, and we will look at this first.

Cross-Sectional Studies

A META-ANALYSIS OF TWIN STUDIES

Our task is made much simpler by the fact that a recent, thorough review of a good part of this evidence already exists, the meta-analysis of twin studies by McCartney et al. (1990) from which we have already reported some data in chapter 3. These authors reviewed a total of some 103 twin studies published between 1967 and 1985, obtaining intraclass correlations for MZ and DZ twins for a number of intellectual and personality variables. The personality variables were grouped in eight trait categories, each represented in from 5 to 20 studies. The average age of the twins in each study was also obtained. It was then a straightforward matter for McCartney et al. to calculate a correlation coefficient across studies between the size of the intraclass correlation and the average age of the twins in the study. This was done separately within each of the eight trait categories, and for MZ and DZ twins. If these correlations are positive, it means that the older twins are more alike. If these correlations are negative, it means that older twins are less alike. In the latter case, if the time course of individual development is substantially determined by genotype, divergence should be less for MZ than for DZ twins.

Table 5.1 summarizes the evidence. The correlations are predominantly negative for both MZ and DZ twins. The overall averages of these correlations were –.30 and –.32 for MZs and DZs, respectively. We can see why the authors titled their article "Growing Up and Growing Apart."

There are some exceptional cases (e.g., Dominance), but it should be kept in mind that the correlations in the table are calculated across studies, so that the *N* on which they are based is as small as 5 in three cases—Dominance is one of these cases—and therefore we should expect a great deal of sampling fluctuation. On the whole, it is probably safer to rely on general trends in the data than to spend a great deal of time speculating about individual traits. The general trend is a predominance of negative correlations.

What are we to make of the fact that the negative correlations between twin resemblance and age are on the whole comparable in size for MZ and DZ pairs?

The most straightforward interpretation would seem to be that the age change, the "growing apart," is mostly environmental in origin. That is, it represents a shift from a relatively greater degree of shared family environmental influences in the earlier years to more and more individually distinctive environments as the twins get older. Had the individual course of development been strongly determined by the genotype, MZs would have been expected to diverge less, DZs more.

TABLE 5.1 Correlations Between Age and Twin Resemblance for Various Personality Traits

Trait	*No. of Studies*	*Age &* r_{MZ}	*Age &* r_{DZ}	*Median Age*	*Age Range*
Activity-impulsivity	14	–.48	–.33	7.6	1-50
Aggression	8	–.09	–.06	11.5	7-49
Anxiety	5	–.34	–.49	20.4	7-30
Dominance	5	.67	.07	30.0	7-50
Emotionality	8	–.11	.30	6.3	1-50
Masculinity-femininity	7	–.81	–.74	16.0	7-50
Sociability	20	–.24	.26	16.5	3-50
Task orientation	5	–.69	–.89	28.0	1-50
Mean		–.30	–.32		

SOURCE: McCartney, Harris, & Bernieri (1990). Copyright American Psychological Association, adapted by permission.
NOTE: DZ *r*s for Activity-impulsivity, Emotionality, and Sociability based on one fewer study. Ages in years. Means calculated via Fisher *z*-transforms.

We already know that both genes and environments contribute to individual differences in traits like these, as indicated by the massive evidence from twin, adoption, and family studies reviewed in the preceding chapters. Indeed, this same set of twin studies fully supports such a finding: Across the eight trait areas in the original tabulation by McCartney et al. (1990), the average MZ twin correlations range from .42 to .59, with a mean of .51, and no overlap at all with the average DZ correlations, which range from .16 to .33, with a mean of .22. What the data in Table 5.1 suggest is not that the genes are unimportant, nor that the environment is, but rather that the growing apart of twins as they grow older reflects a shift of environmental influences from the shared to the unshared category. It is worth noticing that this interpretation suggests a certain transitoriness of environmental effects.

OTHER STUDIES OF ADULT TWINS

Eaves and Martin and their colleagues have reported several behavior genetic analyses focused on age changes in the genetic and environmental determinants of personality (see Eaves et al., 1989, chapter 7). Several of these analyses are based on samples not included in the McCartney et al. (1990) meta-analysis, and thus deserve separate notice here. One caution: The studies are cross-sectional, based on testing twins of varying ages on a single occasion. Thus age differences are confounded with cohort differences. What are interpreted as age changes may sometimes instead reflect changing times. This would presumably only affect conclusions about environmental effects—noticeable genetic change in a population would not be expected within a period of a single generation.

In one twin study of age changes, Eaves and Eysenck (1976b) correlated age with the absolute differences between pair members on an Eysenck Neuroticism scale in a British volunteer adult twin sample. For 402 MZ twin pairs, this correlation was essentially zero (–.02), but for 212 DZ pairs it was statistically significant (+.19). The latter correlation, indicating greater differences for older pairs, is in agreement with the finding by McCartney et al. (1990) of twins growing older and growing apart. The absence of correlation for MZs (and for either group for Extraversion—see Eaves et al., 1989, p. 192), however, is not.

A similar analysis was carried out on data from an Australian adult twin sample (Martin & Jardine, 1986). In five subsamples—MZ and DZ male and female pairs and DZ opposite-sex pairs—the EPQ Extraversion scale yielded no significant correlations of age with absolute pair differences (*r*s in the range –.04 to +.07). Sample sizes were from 351 to 1,233 pairs, and there was an ample adult age range (18-88, with means of about 34 and SDs of about 14), so that any substantial correlation should have been detectable. Most of the subgroups showed an absence of correlation for EPQ Neuroticism as well (*r*s of +.01 to +.02), but in one group, female DZs, there was a statistically significant correlation of +.12 between absolute pair difference and age. As the British DZ twin group had been mostly female, this constitutes a replication of sorts of the British result.

Rose (1988) has reported data for a sample of adolescent and adult twin pairs in Indiana, based on nine factor scale scores from the Minnesota Multiphasic Personality Inventory (MMPI). In general, there were opposite age effects for MZ twins in the two sexes. The older of the 140 female MZ pairs tended to be less alike (seven of nine scales, three of them statistically significant). The older of the 88 male MZ pairs tended on the whole to be more alike—eight of nine scales, although only one was statistically significant. A comparable analysis was not reported for DZ pairs.

What of the two other large adult twin samples, the Swedish and Finnish studies? One, the Swedish study, was included in the McCartney et al. (1990) meta-analysis, the other, the Finnish study, was not. For both, correlations are given by age groups in Table 5.2.

In the Swedish study (Floderus-Myrhed et al., 1980), the overall trends are toward lower correlations as twins grow older. In general, these trends seem more clearly marked for Neuroticism than for Extraversion, and for females than for males. For Extraversion, most of what drop-off in correlation there is appears to lie between the youngest group and the rest.

In the Finnish data, the correlations are less consistent; there is little evidence for generally declining correlations with age for any group, however. This impression is borne out by overall correlations between absolute pair differences and age in this study, which range from –.02 to +.04 across the four gender and zygosity subgroups (Rose & Kaprio, 1988). Note that twins aged 16-24, included in the Swedish data, are not included here. This is an age group which would presumably include the transition from living together to living apart for many pairs. In fact, Rose and Kaprio showed that the degree of separation between MZ twins (on a scale from living together to rarely being in contact) is correlated with differences in Neuroticism scores for both sexes. As was noted in chapter 2, Extraversion is too, for female, but not male MZ pairs. The relationship between separation and personality differences, though it exists, is weaker among DZ than among MZ pairs, perhaps because environmental effects may be obscured by within-pair genetic differences in DZs.

In any case, an association of separation with twin resemblance is causally ambiguous: Is it the twins who are more similar who stay together, and those less similar who elect to live apart, or does the separation itself lead to a divergence of personality due to a decrease in shared environmental factors? Rose and his associates (Kaprio, Koskenvuo, & Rose, 1990) have addressed this question by studying a group of MZ twins from the Finnish study who were living together

TABLE 5.2 Adult Twin Correlations by Age Groups—Scandinavian Studies

	Neuroticism				*Extraversion*			
Ages	*MZf*	*MZm*	*DZf*	*DZm*	*MZf*	*MZm*	*DZf*	*DZm*
				Sweden				
16-28	.62	.53	.29	.24	.58	.51	.27	.18
29-38	.51	.39	.24	.19	.50	.37	.16	.19
39-48	.42	.40	.18	.16	.51	.46	.14	.21
				Finland				
24-29	.46	.31	.21	.18	.47	.47	.17	.18
30-34	.33	.36	.18	.04	.47	.44	.12	.14
35-39	.41	.32	.16	.07	.49	.46	.20	.09
40-44	.45	.34	.16	.15	.56	.56	.04	.18
45-49	.46	.25	.17	.14	.52	.28	.12	.14

SOURCES: Floderus-Myrhed, Pedersen, & Rasmuson (1980); Rose & Kaprio (1988).
NOTE: Pairs per age group, Swedish study, 507-1,240 for MZ groups, 814-1,798 for DZ groups. Finnish study, *N*s by age group not given, but total pairs 1,293 MZf, 1,027 MZm, 2,520 DZf, 2,304 DZm.

in 1975 when initially tested. They were tested again 6 years later in 1981, when some were still living together, some separated but in frequent contact, and some rarely in contact. Correlations for Neuroticism and Extraversion for these three groups are shown in Table 5.3.

In 1981, for Neuroticism in both sexes and for Extraversion in females, separation was associated with personality differences (as was true for the whole sample, as noted earlier). Twins living together were most alike; twins in rare contact were least alike. Was separation the cause or effect of differences in twin resemblance? The answer turns out not to be the same for the two traits. In the case of Neuroticism, the separation appears to have caused the differences in resemblance: The groups who differed in 1981 had been equally alike when they were living together 6 years earlier. In the case of Extraversion, the personality difference appears to have caused the separation. The groups who were different in 1981 had already been different when they were living together in 1975.

Not all questions are answered here, obviously. The samples in several of the subgroups are quite small—as few as 27 pairs in one—with concomitantly large sampling errors. Similar analyses for other

TABLE 5.3 Twin Correlations for Finnish MZs Living Together in 1975 but with Varying Degrees of Separation in 1981

	Neuroticism				*Extraversion*			
	1975		*1981*		*1975*		*1981*	
Status in 1981	*MZf*	*MZm*	*MZf*	*MZm*	*MZf*	*MZm*	*MZf*	*MZm*
Living together	.63	.36	.71	.52	.70	.71	.73	.41
Frequent contact	.58	.50	.41	.43	.59	.52	.40	.48
Rare contact	.65	.52	.37	.24	.43	.40	.26	.27

SOURCE: Kaprio, Koskenvuo, & Rose (1990).
NOTE: Number of pairs: Living together 53-60, Frequent contact 136-178, Rare contact 27-35.

personality traits and for DZ pairs would be of interest. As it is, amount of contact is not a quantitatively powerful predictor of personality resemblance for individual twin pairs. The correlation among MZs in the Finnish sample between absolute difference in Neuroticism score and degree of separation was only +.11; for DZs or for Extraversion, it was less (Rose, Koskenvuo, Kaprio, Sarna, & Langinvainio, 1988).

In summary, a number of the studies we have reviewed suggest that the effects of variation over time in shared environment may be greater for emotional stability than for extraversion. Some studies suggest a stronger effect in females than in males. For MZ versus DZ twins, the results have been inconsistent. There is ongoing controversy concerning the interpretation of the data that are available (Lykken, McGue, Bouchard, & Tellegen, 1990; Rose & Kaprio, with Williams, Viken, & Obremski, 1990). Further data and further analysis are needed. Nonetheless, one can readily agree with Kaprio and his colleagues when they say, "longitudinal data from population-based twin cohorts will provide rich material for these analyses" (1990, p. 275).

AGE DIFFERENCES AMONG RELATIVES

Many traits show systematic age trends. For example, both Neuroticism and Extraversion showed modest tendencies to decrease with age—correlations with age of around –.14 in both males and females in the Australian twin study (Martin & Jardine, 1986). Relationships of age with personality are also found in adoption samples. Significant linear and nonlinear relationships of age with composites of trait ratings were found for children in the Texas Adoption Project (Loehlin, Horn, & Willerman, 1990).

Systematic relationships of this kind imply that relatives more different in age will tend to be more different in personality. That is, parent and child will be less alike than siblings, and siblings different in age will be less alike than those similar in age, and siblings less alike than DZ twins, even though all four groups share, on the average, half their genes. Such a process could be genetic (if expression of the genotype differs with age), or environmental (if individuals more alike in age share more environmental factors), or both.

In the path models of chapters 2 and 3, we allowed for the possibility of a correlation, t, between genotypes of different ages—in particular, from child's to parent's age—but found that it could not be shown to differ significantly from 1.0. This suggests that over the (mostly) adolescent to adult ages covered by these data, genetic age changes were not a major quantitative factor. (This story might be different during childhood.) For environment, the best-fitting models had a shared environment component that was zero for parent and child, but not for siblings, suggesting some role of age differences in the environmental component of the variance.

Eaves and his colleagues (1989) have suggested a specific model for age changes, in which the covariance between two relatives decreases as an exponential function of the age difference between them. Fits of such a model to British familial data on Extraversion and Neuroticism suggested that such effects were difficult to disentangle from nonadditive genetic effects. In models where both were permitted, Extraversion showed practically no drop-off in correlation as age differences increased, and Neuroticism not much (p. 197).

Thus, although age differences among relatives can in theory complicate heredity-environmental analyses, in practice they seem not to be a major quantitative factor, at least for personality traits during the bulk of the age span after childhood.

OLDER TWINS

There is much less information available about how genetic and environmental variation contribute to personality variation during the later stages of the life span. During the past few years, however, initial reports have been forthcoming from a currently ongoing study of older twins, the Swedish Adoption/Twin Study of Aging (SATSA; Plomin & McClearn, 1990). So far the data are largely cross-sectional, from an initial wave of testing of the twins. A longitudinal follow-up is under way. A unique feature of the study is the inclusion of a fairly large subgroup of twins who were reared apart in separate families during all or part of their childhood years.

Data from this study have already been cited at various points in this book in connection with the analyses of particular traits. Here we wish to use the data to address the age question: Are the contributions of genes and shared environments different among older persons than they are among the children, adolescents, and young and middle-aged adults that have served as the subject populations in the bulk of the studies reviewed in this book? The SATSA twins at the initial testing averaged 58.6 years of age, with a fairly wide range of variation, from about 2% below 30 to about 2% above 80 (Plomin, Pedersen, McClearn, Nesselroade, & Bergeman, 1988). Almost 90% were past 40, about 70% above 50, and nearly half were 60 years of age or older.

Table 5.4 shows estimates of total genetic effects (additive plus nonadditive) and shared environmental effects for a number of personality scales included in the initial SATSA testing. These estimates were obtained via model-fitting analyses in the same general spirit as those that have been reported in previous chapters of this book. These are not huge samples (82 to 220 pairs), so that one would expect some sampling fluctuation, and thus should not overinterpret the results for particular traits; nevertheless, it may still be useful to compare the overall median estimates of .30 and .08 for this predominantly older group with the model-fitting estimates arrived at in chapter 3 for the Big Five factors, based on a range of samples that includes this one at its upper end, but are predominantly younger. Table 3.20 on page 67 summarizes those analyses. If we take the second model from that

TABLE 5.4 Estimates of Genetic and Shared Environmental Variance for Personality Traits in Older Twins, Swedish Adoption/Twin Study on Aging

Trait	*Questionnaire*	*Genes*	*Shared environment*
Extraversion	Short form EPI	.41	.07
Neuroticism	Short form EPI	.31	.10
Impulsivity	KSP	.45	.00
Monotony avoidance	KSP	.23	.05
Emotionality	EAS	.39	.06
Fear	EAS	.39	.04
Anger	EAS	.28	.12
Activity	EAS	.23	.12
Sociability	EAS	.24	.13
Openness	NEO	.41	.06
Conscientiousness	NEO	.29	.11
Agreeableness	NEO	.12	.21
Median		.30	.08

SOURCES: Plomin & McClearn (1990). Copyright Academic Press, Inc., adapted by permission. *N*s from Pedersen, Plomin, McClearn, & Friberg (1988); Bergeman & Chipuer (1989).
NOTE: EPI = Eysenck Personality Inventory; KSP = Karolinska Scales of Personality; EAS = Buss & Plomin EAS Temperament Survey; NEO = Costa & McCrae NEO Personality Inventory. *N*s: 82-95 MZ and 171-220 DZ pairs reared apart, and 132-151 MZ and 167-204 DZ pairs reared together. Age: M = 58.6, SD = 13.6.

table (h^2, i^2, c_S^2) as most nearly comparable to the models fit in the SATSA studies, combine the two genetic components, and obtain medians for the Big Five traits, the values of these medians are .41 and .07 for genetic and shared environmental effects, respectively.

Overall, then, the median of .30 for genetic effects in older twins appears to be somewhat less than the .41 found with predominantly younger samples. The effects of shared environment at .08 and .07 appear to be much alike. These conclusions must remain tentative pending their confirmation in a greater range of samples at older ages. Nevertheless, on present evidence, it would appear that an increased influence of environmental events not shared with the twin may be acting to decrease gene-based resemblance during the later portions of the life span, even as the tendency toward diminishing effects of shared environment occurring earlier has pretty much leveled off.

NEWBORN TWINS

How do things start out? There has been at least one study of temperament using newborn twins (Riese, 1990). In this study, part of the Louisville Twin Study (of which more later), full-term twins were observed during their first week of life (average age, 3.7 days); preterm twins and those with medical complications were observed when medically stabilized, just prior to their discharge from the hospital (average age,16.9 days). A series of standardized procedures and observations were undertaken during one feeding-to-feeding cycle. These included such things as the infant's sleeping and waking activity patterns, how attentive it was to various stimuli, how easily irritated and how readily soothed it was, how cuddly and responsive, and so on. These ratings were consolidated into four temperament composites, plus overall sleeping and waking activity. Correlations on these were obtained for MZ twins, for same-sex DZ pairs, and for opposite-sex DZ pairs. The results are shown in Table 5.5.

It is evident from these correlations that there was no tendency for MZ pairs to be more similar than DZ pairs—if anything, one might argue the reverse, although with these sample sizes most of these differences should probably be regarded as simply due to chance. There is a moderate degree of twin resemblance overall—all the correlations in the table are positive. Something is making twins alike. But whatever it is looks more environmental than genetic, in that it operates at least as strongly for DZ as for MZ pairs. Riese suggests that such environmental factors as prenatal and birth conditions may play a dominant role in accounting for the similarity within pairs and the differences between them among these newborn twins. There was some support for this interpretation in the presence of modest but significant correlations between differences on some of the temperament composites and such variables as birth weight and Apgar score (a measure of the physiological status of the newborn).

In the beginning, then, the rudiments of personality/temperament, the individual differences in behavioral patterns among newborns, appear to reflect primarily shared or idiosyncratic environmental factors, not genetic differences. The genes, of course, are playing a major role in regulating the unfolding of behavioral and physiological

TABLE 5.5 Twin Correlations from Neonatal Temperament Assessment

Composite rating	*MZ twins*	*SS DZs*	*OS DZs*
Irritability	.30	.45	.28
Soothability	.06	.59	.30
Responsiveness to stimuli	.12	.06	.13
Reinforcement value	.31	.49	.30
Activity: awake	.09	.12	.16
Activity: asleep	.22	.21	.19

SOURCE: Riese (1990). Copyright Society for Research in Child Development, adapted by permission.
NOTE: Number of pairs—MZ, 45-47; SS (same-sex) DZ, 38-39; OS (opposite-sex) DZ, 70-72.

characteristics, but as a source of individual differences in temperament, they are apparently not a major contributor—or, more accurately, not yet.

THE FIRST YEAR

There have been, by now, a number of studies in which MZ and DZ twins during the first year of life have been rated on various behaviors presumably related to personality/temperament. Many of these studies are individually fairly small, but the central tendency of their results may still be informative.

Some representative results are shown in Table 5.6. This table gives median MZ and DZ correlations over sets of questionnaire scales, rating factors, or the like, and may thus be seen as indicating typical levels of twin resemblance on such measures. The main point of interest is that typical MZ correlations are higher than typical DZ correlations for 9 out of the 10 sets in the table. Thus within the first year, we begin to see some degree of the greater resemblance of MZ than DZ twins that is characteristic of twin data at later ages and may reflect the emergence of a genetic contribution to individual differences in this realm. The differences are not large—the median value of the differences in Table 5.6 is about .12. Doubling that figure would suggest a heritability in the neighborhood of .24. As we have seen earlier, this requires assumptions that are probably not generally

TABLE 5.6 Twin Resemblance During the First Year of Life—Median MZ and DZ Correlations

			Median r		Pairs[a]	
Study location	*Age in months*	*Scales used*	*MZ*	*DZ*	*MZ*	*DZ*
Denver	9	6 IBQ scales	.73	.52	29	61
	9	10 lab composites	.32	.22	35	35
Louisville	3	5 IBR factors	.31	.26	76	51
	6	5 IBR factors	.40	.16	91	54
	9	5 IBR factors	.35	.22	72	35
	12	5 IBR factors	.43	.28	94	46
Louisville	9	4 lab ratings	.54	.42	30	28
	12	4 lab ratings	.42	.51	30	28
U.S.	8	11 tester ratings	.42	.30	117	213
Taiwan	6	9 Carey scales	.80	.60	44	18

SOURCES: Denver—Goldsmith & Campos (1986); Louisville—Matheny (1980), Wilson & Matheny (1986); U.S.—Goldsmith & Gottesman (1981); Taiwan—Chen, Yu, Wang, Tong, & Tien (1989).
NOTE: IBQ = Rothbart's Infant Behavior Questionnaire, filled out by parents; IBR = Bayley's Infant Behavior Record, filled out by tester.
a. Sometimes approximate—can vary slightly for different measures.

defensible, but pretty clearly this is a lower figure than the .60 or so we would have got by doing the same thing with adult twins.

Most of the behavior genetic data at ages as young as these has been obtained from twins, but there has been one adoption study, the Colorado Adoption Project (Plomin & DeFries, 1985), that has examined biological and adoptive sibling and parent-offspring resemblance when the infants were 12 months of age. Sibling data on the Infant Behavior Record, with both siblings measured at one year of age, were consistent with the twin data in suggesting modest heritabilities (Braungart, Plomin, DeFries, & Fulker, in press). The median correlation for 101 biologically related sibling pairs over three IBR factors was .20, and the median for 83 unrelated pairs was essentially nil (actually, –.09). There was only the merest hint of parent-offspring similarity, however (Plomin & DeFries, 1985); the overall median of 48 correlations between adult measures of Neuroticism and Extraversion and presumptively related infant measures was around .03 or .04. What faint shadow of resemblance there was appeared to be environmental—in adoptive and control families, positive correlations tended slightly to outnumber negative ones, whereas among birth parents and offspring, positive and negative correlations were virtually evenly split.

THE SECOND YEAR

The twin data available at this age are primarily from the Louisville Twin Study. Some representative correlations are shown in Table 5.7.

Again, these are averages (medians) over the sets of scales used, to give some idea of typical levels of resemblance. Comparison with corresponding values from the first year (Table 5.6) suggests that in the second year the MZ-DZ differences have increased appreciably—they have risen from a median value of about .12 to one of about .27, which approaches the average MZ-DZ difference of .29 found in the McCartney et al. (1990) meta-analysis of twin studies described earlier in this chapter.

The sibling adoption data from the Colorado study (Braungart et al., in press) again suggests moderate heritabilities at 24 months. The median correlation over three IBR factors was .24 for 88 biological sibling pairs, and .01 for 78 adoptive pairs. In the Colorado data there is still only the same minimal parent-offspring resemblance at 24 months as was found at 12 months (Plomin & DeFries, 1985). Again, there are a few more positive than negative correlations, but not more than at 12 months. The overall median r over the same 48 correlations is a bare .02. Only the control families show any appreciable predominance of positive rs. The correlations between children and birth or adopted parents are about evenly split between positive and negative.

Thus if we take the twin and sibling data to indicate an emergence of genetic influence on temperament during the second year, we must assume that the genes involved (or their effects) are different from those that affect the adult traits or that the measures at the two ages are nonequivalent. A recent report in which child measures (parent ratings) are available up to age 7, suggests that such a situation may persist well into childhood (Plomin, Coon, Carey, DeFries, & Fulker, 1991).

TABLE 5.7 Twin Resemblance During the Second Year of Life—Median MZ and DZ Correlations

			Median r		Pairs[a]	
Study location	*Age in months*	*Scales used*	*MZ*	*DZ*	*MZ*	*DZ*
Louisville	18	5 IBR factors	.43	.14	79	43
	24	5 IBR factors	.48	.20	80	50
Louisville	18	4 lab ratings	.69	.42	30	28
	24	4 lab ratings	.66	.45	30	28

SOURCES: Louisville—Matheny (1980), Wilson & Matheny (1986).
NOTE: IBR = Bayley's Infant Behavior Record, filled out by tester.
a. Sometimes approximate—can vary slightly for different measures.

Longitudinal Studies

YOUNG TWINS

The preeminent longitudinal study of twins in early life is the Louisville Twin Study, which we have encountered several times already in this chapter. So far, we have dealt with the data cross-sectionally, i.e., as indicating twin resemblance in temperament at different ages, but now we wish to look at temperament/personality change as such and examine its genetic and environmental determinants.

Note that accounting for individual differences at a given age and accounting for changes between two ages are different problems, and can have different answers—certainly in a quantitative sense. It is perfectly possible, for example, for a trait to have, say, heritabilities of .60 at two ages, and yet for individual changes in score between those ages to have a heritability of .00—to be entirely due to environmental influences. Indeed, if genes were fixed and unchanging in their effects, that is the *only* way it could be. Genes are not fixed and unchanging in their effects, of course, which makes matters more interesting and allows us to meaningfully ask the question: How much do changes in the expression of genes contribute to personality change?

As we noted at the beginning of this chapter, behavior genetic questions about change are answered in just the same way as behavior

TABLE 5.8 Correlations of MZ and DZ Patterns of Change Across Age—Louisville Twin Study

Factor	*Ages (Months)*	r_{MZ}	r_{DZ}
Lab rating: Emotional tone	9-12	.50	.48
	12-18	.63	.51
	18-24	.80	.41
	12-18-24	.73	.47
IBR: Task orientation	6-12-18	.53	.18
	12-18-24	.49	.21
Affect-extraversion	6-12-18	.35	.06
	12-18-24	.37	.12
Activity	6-12-18	.27	.19
	12-18-24	.52	.18
BSQ: Tractability	36-48	.50	–.05
Lab rating: Surgency	36-48	.35	.08
Median		.50	.18

SOURCES: Wilson & Matheny (1986); Matheny (1983, 1987).
NOTE: IBR = Infant Behavior Record (tester ratings); BSQ = Behavioral Style Questionnaire (parent ratings). All variables are factors based on a number of more specific ratings. Where more than two ages are given, the correlation reflects the similarity of patterns of change for the twin pair. Number of pairs—Emotional tone, 27-33 MZ, 25-31 DZ; 3 IBR factors, 60-66 MZ, 35-40 DZ; Tractability, Surgency, 22 MZ, 23 DZ.

genetic questions about scores at a given age, except that the unit of analysis is a difference score or a profile of changes over time, rather than a score on a single occasion.

Table 5.8 gives some correlations between twin differences and patterns of change for MZ and DZ pairs from the Louisville Twin Study. The sample sizes are small—most are in the range of 20 to 40 pairs—so we cannot place great confidence in the stability of individual correlations. Nevertheless, there is clearly an overall tendency for MZ pairs to show a greater resemblance in their changes than DZ pairs. The median value across the 12 comparisons in the table is .50 for MZs and .18 for DZs, numbers quite similar to typical twin correlations for single traits (the meta-analysis by McCartney et al. [1990] yielded average correlations of .51 and .22, respectively). When we allow for the greater unreliabilities of change scores and take into account the fact that some of the values in Table 5.8 are from the first year of life, when trait heritabilities are generally lower, it would appear that changes in temperament traits during this period are at least as much under the influence of genes as is individual variation at given ages.

ADOLESCENT AND ADULT TWINS

There have been at least three repeated-measures studies of twins in late-adolescent or adult years. In one, Dworkin, Burke, Maher, and Gottesman (1976, 1977) readministered the MMPI and the CPI to 25 MZ and 17 DZ pairs who as high school students had been given these questionnaires in a twin study by Gottesman (1966) some 12 years earlier. (The average ages at first and second testing were 16 and 28 years, respectively.) For 5 of 15 MMPI scales and for 2 of 18 CPI scales, some evidence was found of genetic influences on change. Both instruments showed a statistically significant influence of the genes on changes in average scale elevation. The sample sizes are too small to justify any detailed interpretation of their findings, however.

Eaves and Eysenck (1976a) in Great Britain had much larger samples. They took advantage of the fact that 441 pairs in an adult twin sample had been given different versions of an Eysenck Neuroticism scale on two occasions 2 years apart. These questionnaires had 11 items in common, and Eaves and Eysenck subjected these items to a separate analysis. The basic data are shown in Table 5.9 for two derived variables, change in the total positive response to the items (i.e., change in overall Neuroticism score) and change in the pattern of endorsement of the individual Neuroticism items. (The disproportionate number of female twins is characteristic of volunteer twin samples, although it is more extreme here than in most.)

There is a tendency toward overall positive correlations in the table, except for the opposite-sex pairs, but the tendency is very slight. Most notably, it is no more marked for MZ than for DZ pairs. That is, there seems to be a very weak tendency toward the influence of shared environments on personality change, and there is no evidence that genetic changes contributed to personality change at all during this time period. Of course, 2 years in adulthood might not offer much scope for gene-directed change. Obviously, by far the greatest contributors to change in these data were the factors that do not contribute

TABLE 5.9 Correlations of 2-year Change Scores for Adult Twins, Great Britain

Variable	*MZf*	*MZm*	*DZf*	*DZm*	*DZo*
Neuroticism	.03	.21	.05	.14	–.09
Item pattern	.07	.02	.00	.03	–.01
Pairs	202	51	104	25	59

SOURCE: Eaves & Eysenck (1976a). Correlations calculated from reported mean squares.
NOTE: MZ = monozygotic, DZ = dizygotic, m = male, f = female, o = opposite sex pairs.

to twin resemblance: errors of measurement, unshared experience, and (perhaps) genotype-environment interaction.

Table 5.10 gives data for a sample of young adult twins, most of them initially studied at an average age of 20, while they were students at Indiana University, and followed up and retested 5 years later, at an average age of 25. Again, most of the correlations are positive, suggesting some tendency for twins to share sources of change; and again, the MZ correlations are on the whole no higher than the DZ correlations (medians of .16 and .15), suggesting that these change-producing influences are largely environmental. The Schizophrenia scale may be an exception, but with these modest sample sizes, one would want to see this result replicated before taking it too seriously. And finally, the correlations are fairly low—the bulk of the changes again appear to be idiosyncratic, reflecting errors of measurement and the effects of individual experience.

AN ADOPTION STUDY

In the Texas Adoption Project (Horn, Loehlin, & Willerman, 1979; Loehlin et al., 1981), members of 300 Texas families that had adopted a child through a church-related home for unwed mothers were studied initially when the adopted children were between 3 and 20 years of age, with a median age just under 8 years. All available members of the adoptive families were tested—the adopted child, the adoptive father and mother, and in many families, additional adopted children or biological children of the adopting parents.

TABLE 5.10 Correlations of 5-year Change Scores for Young Adult Twins, Indiana

MMPI Scale	*MZ*	*DZ*
Social maladjustment	.19	–.01
Psychopathic deviate	.07	.22
Schizophrenia	.35	.17
General maladjustment	.14	.09
Depression	.12	.13
Religiosity	.23	.42
Median	.16	.15

SOURCE: Pogue-Geile & Rose (1985).
NOTE: Number of pairs—71 MZ and 62 DZ.

Roughly 10 years later, many of the children in 181 of the families were followed up and retested, now at an average age of about 17, with many of the sample well into young adulthood (Loehlin, Willerman, & Horn, 1987).

At the time of the initial testing, the children were rated by a parent (usually the mother) on a set of 24 bipolar trait scales. These ratings were made again at the time of the follow-up testing. Based on a factor analysis of the initial ratings, three composites of five or six traits each were formed, which were labeled Extraversion, Socialization, and Emotional Stability, and which correspond pretty much to Big Five dimensions I, III, and IV, Surgency, Conscientiousness, and Emotional Stability.

Did the children in a given family tend to show similar changes on these trait composites between the first and second occasions of testing? Did genetically related siblings show more similar changes than adoptive siblings? The answers are given in Table 5.11, which shows correlations between changes in scores on the three composites over the 10-year time span.

The adoptive siblings include two kinds of genetically unrelated pairs, those consisting of two adopted children, and those consisting of an adopted child and a biological child of the adoptive parents. Also shown are correlations based on the small group of biological sibling pairs in the study—small, because not many couples who adopt have two or more biological children of their own.

ABLE 5.11 Correlations of 10-year Change Scores for Adoptive and Biological Siblings, Texas

arent Rating 'omposite	*Adoptive Siblings*	*Biological Siblings*
xtraversion	–.20	–.19
ocialization	–.08	–.19
motional stability	–.08	–.27

OURCE: Loehlin, Horn, & Willerman (1990).
IOTE: Pairings: Adoptive—164 in 115 families. Biological—27 in 18 families.

Did siblings tend to show similar changes? No. The correlations in 'able 5.11 are all negative. Apparently, if one child became, say, more xtraverted over the period in question, his or her sibling tended to e seen by the parent as becoming less extraverted. Did biologically elated siblings show more similar changes than biologically unre-ated siblings, suggesting a genetic contribution to such changes? Again, no. The sample of biologically related siblings is very small, of ourse, so one cannot place much confidence in these numbers, but at ny rate, the biological siblings certainly show no signs of being *more* imilar in their changes than did unrelated children.

Adoptive or biological siblings, who differ in age and may differ in ex as well, might be expected to share fewer environmental factors han do twins, so one might expect less of an effect of shared environ-nent—recall that it was the opposite-sex pairs in the Eaves and Eysenck study (see Table 5.9) who showed negative correlations. But ositive or negative, the correlations are low: As in the twin studies, he bulk of the changes must be classified as due to individual xperience, gene-environment interaction, or errors of measurement.

CHANGES IN GROUP MEANS

One way in which one can examine overall trends in the face of local variation is to look at group means. For adopted children, the birth parents and the rearing parents typically show substantial differences n some aspects of personality. Thus it becomes informative to observe rends over time in the means of adopted children. Do they tend to change in the direction of the birth parents or in the direction of the adoptive parents? The former would suggest genetic influences; the latter, environmental influences.

In the Texas Adoption Project, the birth mothers and the adoptive parents of the adopted children received the MMPI. On several scales indicative of emotional maladjustment, such as Depression, Psychasthenia, Paranoia, and Schizophrenia, the birth mothers had mean scores that exceeded those of the adoptive mothers and fathers by upwards of half a standard deviation (Loehlin et al., 1982). On a scale indicative of poor socialization, Psychopathic Deviate, the birth mothers' means were over a full standard deviation higher. What, then, happens over time to the adopted children's mean ratings on the Emotional Stability and Socialization composites? Table 5.12 shows these results, with those from the biological children of the adoptive parents for comparison.

At the time of the initial testing, there was little difference between the adopted and the biological children in their parents' eyes—what small differences there were tended to favor the adopted children. Between the first and the follow-up testings, the biological children changed little: On the average, their ratings improved on both Socialization and Emotional Stability, but with these sample sizes, neither change was statistically significant. The means of the adopted children, over the same time period, shifted significantly in the direction of poorer socialization and less emotional stability. That is to say, the adopted children changed, on the average, in the direction of their birth mothers. (Probably in the direction of their birth fathers, too, although no test scores were available for the fathers.) A subsequent path model analysis estimated the relative shift on the underlying latent variables to be about four-tenths of a standard deviation (Loehlin, Horn, & Willerman, 1990).

One should not exaggerate this effect. All eight of the means in Table 5.12 lie above the neutral point of the scales: The parents rated both their adopted and their biological children as above average in adjustment and socialization on both occasions. But they did come to regard the behavior of the adopted children less favorably over time. Could it have been solely a change in the attitude of the parents rather than in the behavior of the children? One cannot completely rule out such

TABLE 5.12 Means on Parent Rating Composites for Adopted and Biological Children in Texas Adoption Study at Original Testing and 10-year Follow-up

	Adopted Children		*Biological Children*	
Rating Composite	*Original*	*Follow-up*	*Original*	*Follow-up*
Socialization	33.2	31.6	32.4	33.9
Emotional Stability	28.3	27.2	28.1	28.8

SOURCE: Loehlin, Willerman, & Horn (1987).
NOTE: *N*s—Adopted, 235-243; Biological, 83-85. Midscale (= "average") rating: 30 for Socialization, 25 for Emotional Stability.

a possibility, but in view of the fact that the attitude change developed subsequent to the time of the initial testing, when the parents tended, if anything, to think more highly of the adopted than of the biological children, it seems more likely that a change in the children's behavior was the factor immediately responsible. If so, the shift was in the direction of the children's biological parents, suggesting that it reflected an expression of genes.

Conclusions

In this chapter, we considered age differences in the effects of genes and environments on personality, and genetic and environmental contributions to personality change. We looked at studies of newborns, infants and toddlers, and old age, as well as a summary of studies across the bulk of the life span. We found little evidence of heritability of temperament at birth, evidence of some but not much heritability during the first year, and heritabilities during the second year increasing toward levels characteristic of the major portion of the life span, with perhaps a slight decrease in old age. The tendency for twins to grow less similar as they grow older appeared during much of the life span to be due to a decrease in the effect of shared environments, with a concomitant increase in the contribution of unshared environments. During the later years, the decrease in shared environmental variance may level off, although the increase in the unshared component appears to continue.

Longitudinal studies of personality change in childhood indicated an appreciable genetic contribution to such changes; studies in adolescence and adulthood suggested that most of the observed changes at these ages were due to factors in the unshared environment or errors of measurement—although occasional evidence of genetic effects was reported. In an adoption study, probable gene-based changes in group means were observed, although most changes for individuals appeared to be environmental in origin.

Complexities and Contexts

In this final chapter, we will consider some of the complexities that so far have not been discussed in detail, such as possible correlations and interactions between the genotype and the environment in their contributions to personality variation. We will also look at how the results of behavior genetic analyses of personality traits are similar to or different from the results of behavior genetic analyses carried out in other trait domains. Finally, we will attempt to place our discussion of personality variation in a broader evolutionary context: What can we say about the genetic roots of human personality in general, as opposed to the individual differences with which we have dealt so far? What is the relationship between a generalized "human nature" and dimensions of individual variation?

Correlation and Interaction of Genes and Environment

Genotype-environment correlation and interaction were introduced briefly in chapter 1. Genotype-environment correlation refers to the fact that genotypes may be *exposed* differentially to environments. Genotype-environment interaction refers to the fact that genotypes may *respond* differentially to environments.

GENOTYPE-ENVIRONMENT CORRELATION

Let us begin by considering the relationship to personality of genotype-environment correlation, the tendency of certain genotypes to be exposed preferentially to certain environments. Recall from chapter 1 that three subvarieties exist, passive, reactive, and active, the first caused by the fact that parents are the source of both genes and environments, the second by the reactions of others to genotypically influenced traits, and the third by the genotype's own active selection of its environment.

If one had measures of genotypes and of relevant environments, the assessment of genotype-environment correlation would be simple—just the calculation of a correlation coefficient. But generally one does not. Nevertheless, there *are* methods of estimating the contribution of a passive genotype-environment correlation to the variance of a trait based on data from adoption studies. At least two such methods have been described (Loehlin & DeFries, 1987). They are both based on differences between biological and adoptive families that derive from the fact that children's genotypes are correlated with the environments their parents provide in biological families but not in adoptive families. One method is based on differences between parent-child correlations in the two kinds of families, the other on the differences between the variances of the children.

It turns out that the method using parent-child correlations is decidedly the more powerful of the two, although the comparison of variances may be feasible in circumstances where the correlational method cannot be used—for example, when information is available on adopted and biological children but not their parents.

Table 6.1 provides some examples from the Texas and Minnesota adoption studies. In each case, Eysenckian Neuroticism and Extraversion scales are available, as well as the three scales from the Extraversion domain, Impulsivity, Dominance, and Sociability, shown to correspond well across data sets in the analysis in chapter 4. In the top part of the table, midparent-child correlations and estimates based on

TABLE 6.1 Estimates of Passive Genotype-Environment Covariance for Five Traits in Two Adoption Samples via Correlations and Standard Deviations

	Texas			*Minnesota*		
Trait	*Adopt. r or SD*	*Biol. r or SD*	*GE covar.*	*Adopt. r or SD*	*Biol. r or SD*	*GE covar.*
	Estimated via midparent-offspring correlations					
Neuroticism	.07	−.10	−.02	.05	.25	.02
Extraversion	.04	.12	.01	−.00	.19	.00
Impulsivity	.06	.09	.00	−.02	.14	−.01
Dominance	.02	.15	.00	.10	.21	.02
Sociability	.03	.18	.01	−.00	.28	.00
	Estimated via child variances					
Neuroticism	10.81	9.71	−.24	4.40	4.42	.01
Extraversion	7.44	7.03	−.12	3.61	3.77	.08
Impulsivity	3.51	3.36	−.09	5.42	5.97	.18
Dominance	5.11	5.19	.03	6.06	5.96	−.03
Sociability	4.23	4.00	−.12	4.71	4.80	.04

SOURCES: Loehlin, Willerman, & Horn (1985), and additional analyses; Scarr, Webber, Weinberg, & Wittig (1981).
NOTE: Adopt. = adopted child; Biol. = biological child. GE covar. = estimate of genotype-environment covariance: $2(r_B - r_A)r_A$ for correlations, $(V_B - V_A)/V_B$ for variances. Texas sample: Neuroticism and Extraversion, Eysenck CPI scales, other scales, TTS. *Ns*, 241-283 adopted, 54-60 biological. Minnesota sample: Neuroticism and Extraversion from EPI, other scales, DPQ. Approximate *Ns*, 183-190 adopted, 242-255 biological.

them are shown; in the bottom part, standard deviations and variance-based estimates are given.

The correlation-based estimates in the top part of the table are obviously all very close to zero, and the trivial departures show no consistency across studies. The variance-based estimates in the bottom part of the table are also centered close to zero (5 are positive, 5 negative). The departures from zero are numerically somewhat larger for the variance-based estimates, but none are even close to being statistically significant, based on *F*-tests of the variances, and there is again no consistency across the two studies. We may regard them as reflecting chance fluctuation and conclude that at least for these traits, passive genotype-environment correlation makes a negligible contribution to trait variation. This conclusion agrees with that of Plomin and his colleagues (Plomin & DeFries, 1985; Plomin, DeFries, & Fulker, 1988), based on comparisons using Colorado Adoption Project adoptive and control families. They examined both parent-offspring correlations and child standard deviations at each year of age from 1 to 4 and concluded that passive genotype-environment correlation did not have an important influence upon their measures of infant and child temperament.

There is nothing about the method that forces this outcome. For intelligence, quite different results have been reported. Estimates of genotype-environment correlation centering in the neighborhood of .15 to .25 were obtained for IQ using these same procedures (Loehlin & DeFries, 1987).

What of the active and reactive types of genotype-environment correlation? The methods just described will not work for these, because active and reactive genotype-environment correlation directly reflect the child's genotype, which does not depend on whether he or she is an adopted child or not. That is, if an emotional child affects his environment in such a way as to enhance or diminish his emotionality, this need not depend on whether he is living with biological parents or living in an adoptive family. To assess these forms of genotype-environment correlation, one might study children in environments that are differentially responsive to their genotype or that offer greater or lesser opportunities to seek out trait-enhancing or trait-diminishing inputs, and examine the variability of the trait in such environments. Alternatively, one might employ a strategy suggested by Plomin et al. (1977) for screening measured environmental features in adoptive families for possible evidence of genotype-environment correlations. The proposal was to correlate aspects of the birth mother's personality with aspects of the adoptive family environment. The method assumes that selective placement of the child with respect to these personality features can be excluded. If so, there would be no reason for the birth mother's personality to be systematically correlated with features of the adoptive family environment *except* via the behavior of the child in provoking or selecting such environmental features. The presence of a correlation of this sort would thus be evidence of the presence of a genotype-environment correlation of the active or reactive variety.

Plomin and DeFries (1985) applied this latter approach to data from the Colorado Adoption Project for 1- and 2-year-old adopted children. And indeed they found some low but significant correlations (in the range .15 to .20) between the personality of the birth mother and measures of the adopted child's environment. For example, the birth mother's Neuroticism score was correlated –.20 with a family environment factor emphasizing cohesion, expressiveness, and independence. If this does reflect a reactive genotype-environment correlation, it would suggest that the family is responding positively to the presence of an emotionally healthy child. This would need to be a pretty subtle effect, because as we saw in the last chapter, direct ratings of the child's behavior at these ages show little correlation with the birth mother's personality traits. Furthermore, this particular correlation decreased to –.12 as sample size increased and was an even more modest –.08 when family environment was assessed again at age 3 (Plomin et al., 1988). Probably at this stage we should regard results of this sort as illustrative of the method rather than as firm evidence of the phenomenon.

An approach of this kind could also be used with twins reared apart. MZ twins should work best, especially if nonadditive genetic effects are important for the trait in question, and again selective placement would need to be excluded. In this strategy, one would correlate a personality trait of Twin A with the environment of Twin B. If these are correlated, it would presumably mean that Twin B's genotype has selected or elicited the feature of the environment in question; that is, it would be evidence of genotype-environment correlation of the active or reactive kind.

Scarr and McCartney (1983) have proposed a developmental theory of genotype-environment correlation. They conjecture that such correlation should primarily be of the passive variety early in a child's development, shifting more to the active form as the child grows older. We have presented some evidence that there is little passive genotype-environment correlation in late adolescence and adulthood, which is consistent with the theory. But the Colorado data suggest that there is little passive genotype-environment correlation in infancy either, and perhaps appreciable reactive and active correlation at early ages. Neither of these latter results is especially supportive of the Scarr-McCartney hypothesis, although they can hardly be said to provide a strong test of it.

GENOTYPE-ENVIRONMENT INTERACTION

What about the other potentially complicating factor in heredity-environment analyses, genotype-environment interaction? Here, particular combinations of genes and environments have idiosyncratic effects. In many of the analyses reported in this book, such interactions will look like random error. At several points, we have noted that the large residual component of variation within families might contain effects of genotype-environment interaction in addition to within-family environmental variation and measurement error.

Because genotype-environment interaction could conceivably take many forms, it is unlikely that an exhaustive analysis of its effects for any trait will be available soon. Nevertheless, beginnings have been made. Plomin and DeFries (1985) in the Colorado adoption study examined a number of potential interactions using birth mothers' traits as a rough index of child's genotype and characteristics of the adoptive parents or adoptive home as environmental factors. Out of a dozen combinations examined at each of two ages, 12 and 24 months, they found contributions to variance ranging from 0% to 2.5%; none differed significantly from zero. They do, however, discuss briefly a couple of cases that approached statistical significance. In one such instance, an adopted child's emotionality reflected the level of its adoptive mother's anger to a greater degree when the birth mother was low in anger than when she was high. Presumably, a genetic predisposition in the child tended to lead to emotionality whenever it was relatively strong, but when it was weak, environmental input from the adoptive mother governed the outcome.

In the Swedish Adoption/Twin Study of Aging, Bergeman, Plomin, McClearn, Pedersen, and Friberg (1988) pursued the search for genotype-environment interactions among 99 sets of identical twins who had been reared apart for at least a substantial portion of their childhood.

Measures of Extraversion, Neuroticism, Impulsivity, and Monotony Avoidance were related to socioeconomic status and to Family Environment Scales (FES) that rated various aspects of the family within which a child had been reared. Basically, Bergeman et al. predicted Twin A's personality score using three predictors: the score of Twin B, a measure of Twin A's family environment, and the interaction of these. Twin B's score was taken as a measure of Twin A's genotype obtained independently of Twin A's family environment. This permitted the interaction between independent measures of the genotype and family environment of Twin A to be evaluated in a way that would be impossible if only a single individual had been involved.

It should be noted that this is a somewhat hazardous procedure if active or reactive genotype-environment correlation is present. For then the purported measures of family environment may in fact be reflecting the individual's genotype, and interactions become difficult to interpret—if indeed they do not vanish altogether.

Nevertheless, Bergeman and her colleagues found a number of significant interactions, accounting for 2% to 6% of the variance of the personality traits involved. Not all of these interactions are easy to interpret, but several were similar to the one described earlier for emotionality. For example, in subjects with a strong genetic predisposition toward extraversion, it did not matter whether the family environment was controlling or not—the individual was extraverted in either case. But with a weak genetic disposition it did matter. For such individuals, growing up in low-controlling families led to more extraversion than growing up in high-controlling families. Thus, for some genotypes (low extraversion), degree of family control mattered; for other genotypes (high extraversion) it did not.

GENETIC AND ENVIRONMENTAL INFLUENCES ON THE PERCEIVED ENVIRONMENT

Two siblings describe their family environment. Their descriptions differ. Does this reflect a genotype-environment correlation, a genotype-environment interaction, both, or neither? It may be that the two siblings were in fact treated differently as a consequence of differing

TABLE 6.2 Correlations of Ratings of Family Environment on Affection and Control Dimensions by Twin and Sibling Pairs

Group	*Reared*	*Affection*	*Control*	*Equation for Model Fitting*
Ohio MZ I	together	.65	.49	$h^2 + i^2 + c^2$
Ohio MZ II	together	.63	.44	$h^2 + i^2 + c^2$
SATSA MZ	together	.66	.60	$h^2 + i^2 + c^2$
SATSA MZ	apart	.37	.00	$h^2 + i^2$
MSTRA MZ	apart	.28	.18	$h^2 + i^2$
Ohio DZ I	together	.19	.46	$\frac{1}{2}h^2 + c^2$
Ohio DZ II	together	.21	.54	$\frac{1}{2}h^2 + c^2$
SATSA DZ	together	.42	.31	$\frac{1}{2}h^2 + c^2$
SATSA DZ	apart	.29	.17	$\frac{1}{2}h^2$
MSTRA DZ	apart	.28	−.02	$\frac{1}{2}h^2$
California SS	together	.45	.62	$\frac{1}{2}h^2 + c^2$
California OS	together	.46	.56	$\frac{1}{2}h^2 + c^2$

SOURCES: Rowe (1981, 1983); Plomin, McClearn, Pedersen, Nesselroade, & Bergeman (1988); Bouchard & McGue (1990).
NOTE: Ohio I—Acceptance-Rejection and Firm vs. Lax Control factors from Shaefer's Children's Reports of Parental Behavior (father and mother combined); 46 MZ, 43 DZ pairs. Ohio II and California (SS = same sex, OS = opposite sex siblings)—Acceptance-Rejection and Restrictiveness-Permissiveness factors from Moos Family Environment Scale (FES); 59 MZ, 31 DZ, 52 SS sibling, 66 OS sibling pairs. SATSA (Swedish Adoption/Twin Study of Aging)—Relationship and System Maintenance factors from FES; 116-130 MZ, 156-174 DZ pairs reared together, 68-79 MZ, 169-179 DZ pairs reared apart. MSTRA (Minnesota Study of Twins Reared Apart)—Cohesion vs. Conflict and Positive Constraint factors from FES; 45 MZ, 26 DZ pairs.

behaviors—a case of reactive genotype-environment correlation. Or it may be that they were treated the same, but interpreted the treatment differently because of differing genetic dispositions—a case of genotype-environment interaction. Or both of these processes may have occurred. Or neither—the siblings might have been treated differently by chance or for reasons unrelated to their genotypes and simply are veridically reporting this fact.

If we take the child's description of his or her family environment as a behavior, we can subject it like any other behavior to a behavior genetic analysis by means of a twin study or other appropriate design and begin to make distinctions among these possibilities. Several studies of this sort have been done. Some representative results are shown in Table 6.2, involving two adolescent Ohio twin samples and a California sibling sample studied by Rowe (1981, 1983), reared-

TABLE 6.3 Model Fitting to the Correlations of Table 6.2

				Estimates		
Factor	χ^2	*df*	*p*	h^2	i^2	c^2
			All samples			
Affection	6.42	9	>.50	.47	.00	.17
Control	20.54	9	<.02	.18	.00	.35
			Adolescent samples only			
Affection	3.62	4	>.30	.55	—	.09
Control	2.23	4	>.50	.00	—	.52

NOTE: Top part: equations and correlations as given in Table 6.2, solved for *h*, *i*, and *c*. Bottom part: SATSA and MSTRA equations omitted, and remainder solved only for *h* and *c*.

together and reared-apart adult twin samples from the Swedish Adoption/Twin Study of Aging (Plomin, McClearn, Pedersen, Nesselroade, & Bergeman, 1988), and reared-apart twins from the Minnesota study (Bouchard & McGue, 1990).

The two Ohio twin samples, predominantly high school students, and the California siblings, aged 11-19, individually filled out questionnaires describing their home environments. The adult twins in the Swedish and Minnesota studies described the childhood family environments in which they had been reared—for twins reared apart, these were two different ones. Results are shown in the form of twin correlations for two factors derived from the scales of the questionnaires, in each case one factor is focused on warmth and affection, the other on discipline and control.

The correlations for the affection factors show a pattern suggestive of genetic influence—MZ correlations generally in excess of DZ and sibling correlations, and appreciable correlations for the separated twins. The correlations for the control factors look much more environmental—DZ and sibling correlations as high as MZ correlations, and little if any correlation for reared-apart twins.

These impressions were tested by model fitting, using the equations shown to the right in Table 6.2. The top part of Table 6.3 shows the results. The model shows a good fit to the correlations for the Affection factor; a less satisfactory fit for Control. For neither factor does nonadditive genetic variance appear important, but the two factors are almost mirror opposites for estimates of additive genes and shared environment: .47 and .17 for Affection; .18 and .35 for Control. In the bottom part of the table is shown the fit of a reduced model to the data of the U.S. adolescent samples alone, excluding the adult twin samples. Now the fit is good for both the Affection and Control factors, and the contrast between them is even more marked: a strong genetic contribution for Affection, a strong family environmental one for Control.

What does this mean? Well, it does *not* mean that in general, family members agree as to what the family level of discipline is, but disagree on the family level of warmth. The overall levels of correlation in the two columns of Table 6.2 are roughly comparable, at a level representing fair, but far from perfect, agreement for both. What the data seem to show is that affection (or the perception of affection) is contingent on gene-based individual differences in a way that discipline (or the perception of discipline) is not. Individuals who are more alike genetically agree more in their judgment of whether their environment is warm and supporting. This is not true for judgments concerning strictness or permissiveness of discipline. Perhaps the simplest interpretation is that a reactive genotype-environment correlation prevails in the former case, but not in the latter. If expressions of affection from parents and siblings depend to a considerable extent on a child's elicitation of them, but the disciplinary standards of a family are impartial, such a pattern of correlation might be observed. It would be most interesting to have data of this kind analyzed jointly with (a) measures of the personalities of the individuals involved, and (b) judgments about the family environment made by others than the affected individuals.

Personality and Other Trait Domains

In surveying the broad landscape of genetic and environmental influences on personality traits, one of the striking initial impressions one gets is that everything tends to look pretty much the same. Among the first four of the Big Five factors described in chapter 3, for example, no estimate under either of the two main models tested differed by more than .10 across the four trait domains (Table 3.20, p. 67). There *are* some differences, but the general consensus is of a moderate additive genetic component, a small shared environmental component (which may decline over time), and a component of intermediate size that might represent either nonadditive effects of genes, a special resemblance of MZ twin environments, or both. And, of course, there is always a substantial residual, typically about half the variance, reflecting measurement error, nonshared environmental effects, and (possibly) genotype-environment interaction.

Is there something about behavior genetic methods (or the forces of biological evolution) that makes this sort of result inevitable? To obtain some perspective, it may be helpful to compare findings from the personality domain with those from other behavioral domains, such as abilities, attitudes, and interests.

Some data from the National Merit Twin Study (Loehlin & Nichols, 1976) provide a convenient summary of twin resemblance across several trait domains. The twins in this study were located among U.S. high school juniors who took the National Merit Scholarship Qualifying Test (NMSQT) in 1965, and about a year later filled out a battery of personality, attitude, and interest questionnaires in a mail survey. Each of the correlations in Table 6.4 represents a median value across the set of scales indicated. It is evident that the correlations between twins tend to be much higher for abilities and activities, say, than for self-concepts. The right-hand columns of the table show two derived indices—h^2, a rough estimate of heritability, obtained as twice the difference between the MZ and the DZ correlations, and c^2, obtained as $r_{MZ} - h^2$. What differs strikingly across the trait domains in the table is not the heritabilities, which vary only in the range .30 to .48, but the estimates of shared environment, which drop downward from .38 for overall ability to −.14 for self-concepts. The extremes provide an interesting contrast: the MZ and DZ correlations for the total NMSQT score and the typical self-concept cluster differ by the same amount, .24, leading to the same estimate of genetic contribution, .48. But the difference in level of correlation leads to a striking difference in the estimate of shared environmental effects: high for those factors leading to academic achievement and low for those factors leading to

TABLE 6.4 Typical Correlations from Several Trait Domains, National Merit Twin Study, and Implied Estimates of Heritability and Shared Environment

		Median r		*Estimate*	
Trait Domain	*No. of Scales*	*MZ*	*DZ*	h^2	c^2
Overall ability	1	.87	.63	.48	.38
Ability subtests	5	.74	.52	.44	.30
Activities	17	.64	.49	.30	.34
Personality scales	27	.50	.28	.44	.06
Ideals, goals, vocational interests	31	.37	.20	.34	.03
Self-concepts	15	.34	.10	.48	–.14

SOURCE: Loehlin & Nichols (1976).
NOTE: Ability—NMSQT total score and subtests; personality—CPI scales; others—clusters from an item cluster analysis. Total *N*s: approximately 514 MZ, 336 DZ pairs. $h^2 = 2(r_{MZ} - r_{DZ})$; $c^2 = (r_{MZ} - h^2)$.

self-definition. Indeed, if we take the negative estimate in the latter case seriously, it suggests that contrasts of the twins with each other, either by the twins themselves or by those around them, may be playing a major role in the formation of their self-concepts.

ABILITY VERSUS PERSONALITY

The higher correlations for ability than for personality obtained in the National Merit Twin Study are typical. McCartney et al.'s (1990) meta-analysis reviewed 42 twin studies containing an IQ measure and reported a mean correlation of .81 for MZ twins and .59 for DZ twins. You may recall that that study found considerably lower average twin correlations of .51 and .22, respectively, across a variety of personality variables. Part of this difference is no doubt due to differences in reliability of measurement; IQ tests tend to have higher reliabilities than personality scales. Another part of the difference may be due to age. The IQ studies were done, on the whole, with younger twins (median of 7.8 years, versus about 16 for personality). But plausible adjustments for the effects of unreliability and age would still leave the IQ correlations higher, and of course, age differences could not explain the similar differences within the National Merit data.

What about changes over time? The McCartney et al. (1990) meta-analysis of twin studies showed twins tending to grow less similar in personality as they grew older, to about the same degree for identical and fraternal twins. Their results for IQ were different. DZ twin resemblance tended to show the same pattern for IQ as for personality: There was a negative correlation of –.25 between the age of twins in a study and the size of the IQ correlation between them. MZ twins showed no such tendency: The equivalent correlation was +.15 (there were some negative correlations for other cognitive traits). Such a pattern would lead to increases in heritability estimates for IQ with age, as well as decreases in estimates of the effect of shared environment.

Changes in the personality traits of individuals over time, at least past early childhood, seem mostly to reflect unsystematic environmental changes. For IQ, however, an appreciable genetic component of change is often reported. Age-to-age changes in IQ in the Louisville twin study tended to be more similar for identical than for fraternal twins (R. S. Wilson, 1978). A model-fitting analysis of IQ data from the Texas Adoption Project suggested that IQ changes over a 10-year span reflected new genetic inputs, although the effects of shared environment were declining (Loehlin, Horn, & Willerman, 1989).

The main point, in any case, is that heredity-environment analyses in another trait domain, ability, show differences from those for typical personality traits, most notably in a substantially larger contribution of shared environment for IQ. The values of c^2 for abilities in Table 6.4 run some .25 to .30 higher than corresponding estimates of c^2 for the Big Five traits in chapter 3. Greater genetic contributions to change may be present in the case of IQ, although overall genetic contributions are not so different for IQ and personality.

Evolution, Human Nature, and Individual Differences

The story so far: Individual differences in personality (and other psychological traits) appear to reflect genetic differences between people. They also reflect environmental differences. A part of the environmental variation is associated with shared family environments, but for personality traits this portion is small and tends to decrease with age, so that for many traits it may be negligible among adults.

For some traits, the genetic portion of the variation appears to be almost entirely due to the simple additive effects of genes; for other traits, nonadditive relationships among genes may play an appreciable role. Moreover, the genes and environment affecting personality may be correlated in several different ways or may interact—although these complications are more often discussed in theory than convincingly demonstrated to be important empirically.

All this has to do with individual differences; however, the genes and environment are also involved in what individuals have in common. Most grand personality theories, such as those of Freud or Jung, have been fully as concerned with the general mechanisms of personality, the human nature that we all share, as with individual variations on that theme (Buss, 1990).

It is important to realize that the twin and adoption studies that we have emphasized throughout this book are focused on what differs among individuals, not on what is shared. The fact that most humans have two legs or five fingers on each hand is highly genetically determined, but a traditional twin or adoption study of characteristics such as these would not reveal this fact. DZ twins are just as similar as MZ twins in these respects, and adopted children resemble their adoptive parents just as much as their biological parents. In fact, if number of legs and fingers are genetically fixed and universal in the species, then environmental sources of variation, ranging from accidents of embryonic development to chain saws, will provide the major source of variation in finger and leg number among humans. A conventional behavior genetic study of one of these characteristics would yield only this information. The study of common human nature is primarily the province of general psychology or evolutionary biology rather than of behavior genetics, although the latter can provide some clues to evolution, as we shall see.

In recent years, there has been increasing interest in looking at human personality from a general evolutionary perspective (e.g., Buss, 1984, 1990, 1991), much of it stemming from E. O. Wilson's work in sociobiology (E. O. Wilson, 1975, 1978). The role of such evolutionary notions has for the most part been to provide a theoretical perspective on personality, rather than to lead to tight, empirically testable hypotheses. Many of the key propositions refer to behavioral expressions today that are related back to natural selection among humans in the environments of prehistory, about which our knowledge must largely be speculative. Nevertheless, even if all that such a perspective could do was call to our notice behaviors of evolutionary importance—such as sexual jealousy or mate choice—or provide new ways of looking at old phenomena—such as Daly and Wilson's (1990) reassessment of the Oedipus Complex—it would be worth our consideration. And as we shall see, such an approach can sometimes do more than that.

PERSONALITY TYPES—ONE, TWO, OR MANY?

An old issue in personality theory is whether human personality comes in distinct subvarieties, *personality types,* or is best conceptualized as varying continuously along dimensions, *personality traits.* Evolutionary biology has something to say on this point. For example, Tooby and Cosmides (1990) argue that the nature of human reproduction makes the emergence of genetically based personality types unlikely. That is because parental genes get reshuffled into new combinations in every generation, so that it becomes extremely unlikely that a system depending on a coherent set of genes could be transmitted as a unit from parent to child. What we should find, argue Tooby and Cosmides, is a general human nature, with variations on it provided by variation in individual genes that affect such things as thresholds of behavioral expression. About the only possibility of true emergent types would be on the basis of single genes that could act as switches between alternate courses of development. And the existence of such alternatives would in itself not be trivial to explain in evolutionary terms—it would require that each human possess the genetic repertoire for both lines of development, and this in turn requires an explanation why one variant has not won out at the expense of the other by having a slight reproductive advantage over evolutionary time. One genetic possibility would be so-called frequency-dependent selection, in which either alternative has a reproductive advantage if it is more rare. For example, if there is a preference for novelty in mate choice, perhaps originating as a mechanism to avoid inbreeding, it would lead individuals to choose as mates the less common of two variants in a population. Such a system would tend to produce an equilibrium: As either variant becomes less common, it obtains a reproductive advantage that leads it to increase in frequency in the population until it is no longer rare, and loses the advantage.

In the absence of an evolutionary perspective, would one have thought to look for evidence of frequency-dependent selection as a justification for postulating personality types?

Tooby and Cosmides suggest that there may be one clear example of genetically based personality types in humans, namely, the two sexes. Here the conditions of reproduction place different biological demands on males and females, and such factors as differing certainty of parentage and different amounts of reproductive effort can be supposed to have produced some divergent behavioral evolution in the two sexes. But even here there is complete reassortment of genes in every generation, so that each individual must have the genetic repertoire for both sexes—the differentiating factor is a simple genetic switch on the Y chromosome that determines one or the other path of development.

Gangestad and Simpson (1990) suggest another possible candidate for personality types, in this case among women, which they label restricted and unrestricted sexuality. Restricted sexuality refers to a requirement of time, commitment, and closeness to a romantic partner before entering a sexual relationship; unrestricted sexuality refers to a willingness to enter sexual relationships with minimal demands for attachment and commitment. Gangestad and Simpson offer several lines of argument as to why these might represent personality types rather than simply poles of a continuous trait. First, they obtained some evidence that this trait has a bimodal distribution—a scale devised to measure the tendency toward unrestricted sexuality had more cases toward the high and low ends than in the middle. Second, they argued that these two strategies take different routes toward reproductive success, and that they ought to be subject to frequency-dependent selection. One strategy benefits from paternal investment, the other from the quality of paternal genes. The idea is that the woman who places fewer demands on prospective mates will enjoy a wider selection, particularly among those males bearing genes facilitating sexual success, whereas the woman who takes the route of restricted sexuality must settle for a male who will meet her greater demands. She can expect to lose out with the most sexually attractive, but offset this with better potential for long-term assistance and support in childrearing.

Gangestad and Simpson argue that these two strategies should be frequency dependent, in that if there are many women who are sexually available, none can expect to have as much success in attracting the most sexually desirable males as if there were only a few. Conversely, if there are many women pursuing a highly restricted mating strategy, each will have less chance of landing the cream of the good provider crop.

Gangestad and Simpson tested an interesting and nonintuitive hypothesis deriving from their theory, namely, that women of the sexually unrestricted personality type ought to have more sons than daughters among their progeny. The logic is this: Males vary more in reproductive rate than females, so if the unrestricted women are to get the most benefit from the quality genes they attract, they should have more sons. In three different samples using different approaches, Gangestad and Simpson found modest support for this hypotheses. Correlations between sex ratio and unrestrictedness were in the range .16 to .19. For example, in one study, women listed in *Who's Who in American Women* who were in occupations characterized by less restricted personality traits (e.g., actresses and marketing or sales) had a higher proportion of male offspring than women in occupations characterized by more restricted personality traits (e.g., educational administrators and bank officers).

A different position was taken by Draper and Belsky (1990), looking at a somewhat similar pair of alternative reproductive strategies. They postulate an environmentally set switch that channels individuals in one direction or the other. They use the population biologist's labels *r*- and *K*-selection for the two strategies. The first, *r*-selection, refers to a strategy of emphasizing reproduction at the expense of parental care, producing many offspring with little investment in each one; the second, *K*-selection, refers to producing just a few offspring with a relatively great investment in the care of each. In their original application in population biology, these strategies referred to species differences, such as those between fish (more *r*-selected) and the great apes (more *K*-selected). Draper and Belsky use the terms to refer to two different mating strategies among humans.

Draper and Belsky propose that father-presence or father-absence during childhood is the environmental switch that leads the child to develop toward the *K* or the *r* pattern of behavior. Father presence is a symbol of a stable environment in which enduring pair-bonding and paternal support will be the rule and reproduction will often be delayed to gain these benefits. Father absence is a symbol of an environment in which sexuality will begin early, females will evaluate males more by their current appearance and status than by their long-term potential, and males will tend to exhibit hypermasculinity, boasting, and risk-taking behaviors.

Draper and Belsky emphasize that every individual has the genetic potential for both of these strategies, and that each is adapted to its circumstances (in the sense of maximizing long-term reproductive success under the given conditions), and thus that the total arrangement confers the reproductive benefit necessary for the system to have evolved.

The *r* and *K* strategies have also been emphasized by Rushton (1985) as descriptive of characteristic genetically based personality differences between human racial groups, with Mongoloid populations having evolved more in the *K* direction and Negroid populations having evolved more in the *r* direction, with Caucasoid populations intermediate. Not surprisingly, Rushton's views have created something of a furor (for a critique of Rushton's position favoring an alternative more along the lines of Draper and Belsky, see Mealey, 1990). Two additional points might be made. First, because the variation among individuals within racial groupings on personality traits is vastly greater than average differences among racial groupings, racial stereotyping is still a mark of ignorance, even if Rushton's views about racial differences turn out to be wholly correct. Second, racial comparisons are one of the more direct ways of testing propositions about human evolution. For those curious about our human history, it would be unfortunate if pressures and prejudices in the political arena should deny the use of this source of insight.

HUMAN NATURE AND HUMAN VARIATION

Suppose that there are no distinctive genetically based personality types; that there is indeed just one human nature plus variations on it. Most of the evidence in this book suggests that an appreciable source of these variations is genetic. How does this fact fit into a general evolutionary framework?

One possibility is that genetically varying traits are ones for which natural selection has been weak or inconsistent—"genetic junk" in Thiessen's (1972) phrase. That is, variations in personality style may not have been strongly enough associated with reproductive success in our evolutionary past for any great restriction in their genetic range to have occurred. In this view, the presence of high heritability may be evidence that a trait is not of much evolutionary importance. Or, alternatively, changing environmental conditions may be in the process of shifting an optimum from one level of a trait to another, leaving genotypes strung out along the way. In either case, the distinction between broad and narrow heritability becomes important. If a trait is subjected to selection pressure, additive genetic variance will tend to decrease, leaving a relatively larger share of nonadditive genetic variation. The presence of appreciable genetic dominance or epistasis for a given trait may be evidence that the trait has been under natural selection. This possibility has especially been emphasized by behavior geneticists in the biometrical tradition (e.g., Jinks & Fulker, 1970; Martin & Jardine, 1986). Thus the choice between models featuring nonadditive genetic variation and those featuring special MZ twin environments becomes of considerable theoretical interest. As we have seen, the data are ambiguous on this point, and therefore studies bearing on this issue (e.g., of twins reared apart) become of special importance.

Why are heritabilities often rather similar across different trait domains? One argument on this point, following Allen (1970), is based on the assumption that natural selection for personality traits is mainly stabilizing, that is, that there tends to be selection against extreme values of the trait at either end. In this view, genetic variation on any trait will tend to accumulate via mutation until it is visible to natural selection against the background of environmental variation present, at which point further variation is selected against. Thus the *relative* amounts of genetic and environmental variation present may not differ much across traits—it is the *absolute* amount of variation that is the index of biological importance: little variation for a biologically critical trait, more for a biologically trivial one (Loehlin & Nichols, 1976). Of course, we must be able to define variation in an independent way in order to evaluate such a hypothesis; our usual process with personality traits of using arbitrary scales or taking as our unit the amount of observed variation (as in standard scores) would not be appropriate.

Quite separate from this issue is the matter of the distribution of environmental variation between shared and unshared portions, which as we have seen may vary considerably over age and across different trait domains. In addition, we might expect this to differ under conditions of *r* and *K* selection, with shared family environment playing a greater role in the latter.

EVOLUTIONARY PERSONALITY PSYCHOLOGY AND THE BIG FIVE

What relationship, if any, do the Big Five personality dimensions have to an evolutionary view? Buss (1991) suggests that they may define the major dimensions of the "adaptive landscape," in that perceiving and adapting to other people are the crucial tasks in human long-term reproductive fitness. The assertiveness and agreeableness in the first two factors are the key to the formation of social structures ranging from pair-bonds to the coalitions of group and tribe. The conscientiousness and trustworthiness in the third factor and the emotional stability of the fourth are crucial to the endurance of such structures. The role of the fifth factor is less clear, but perhaps a capacity for innovation may be the key. It is also possible that this factor has a somewhat different evolutionary status than the first four; it will be recalled that it shows less evidence of nonadditive genetic variance than the others.

PERSONALITY AND REPRODUCTIVE FITNESS

The genetic basis of present-day personality may chiefly reflect selective forces among our remote ancestors, but one can still be curious about what is going on in this area today. After all, it could affect the kind of people that are going to be around tomorrow. There seems to be remarkably little information about the current relationship of reproductive fitness to personality, but one such study has recently been reported by Eaves, Martin, Heath, Hewitt, and Neale (1990), based on data from the Australian twin sample. Each adult female in the study was asked how many children she had and whether she was still menstruating. There were 1,101 postmenopausal women for whom data on completed reproduction were available, as well as Extraversion and Neuroticism scores from the Eysenck Personality Questionnaire. Thus it was possible to ask, for example, whether women who were high on Extraversion or Neuroticism tended to have more children than women who were low.

The results were that neither Extraversion nor Neuroticism as such was related to reproduction, but that in combination they were. The women with the most offspring were those high on Extraversion and low on Neuroticism, and their polar opposites, the neurotic introverts. Those with the fewest children tended to fall at the other two extremes: the stable introverts and the neurotic extraverts. The pattern is intriguing, though not easy to interpret, and we do not know whether it also holds true for fathers (or whether it is affected by the fact that these women were all Australians, or all twins). It is to be hoped that other researchers in large-scale studies of personality will ask about reproduction, or that researchers in large-scale studies of reproduction will include a personality scale or two, so that we can get a better grasp on selective trends that may currently be going on.

Conclusions

In this chapter, we have looked at a number of complexities and contexts that give added depth to the research findings reviewed in the previous chapters. We began by looking at the complexities implied by the fact that genes and environments may be correlated or may interact. Next, we considered the evidence that there may be genetic influences on how environments are perceived that may complicate inferences from self-report measures. Then we looked at the findings from heredity-environment analyses of personality traits in the context of findings from similar analyses in other trait domains, in particular, intelligence. Finally, we looked at behavior genetic analyses of individual personality variation in the context of a biologically evolved "human nature."

Where do we go from here? First, it is clear that much research has been done on the simple empirical question of the relative contributions of various sources of individual variation—nonadditive and additive genetic effects, shared and unshared environments—for various human personality traits, and for their changes over time. But it is also clear that much research remains to be done. Only a handful of traits have been at all intensively studied. It is not at all obvious how many should be studied—what the optimum level of analysis is. Five traits? Dozens? Hundreds? Or even thousands of individual personality items? Many portions of the life span have received relatively little behavior genetic attention so far, for example, middle childhood and old age. The complexities of genotype-environment correlation and interaction represent a research area which has barely been touched empirically. Obviously, there are many opportunities still lying ahead. It is my own view that such research will gain greatly in relevance if it is informed by a broad biological and evolutionary perspective as well as rich theories of the cultural and physical environments within which human development takes its course. I hope some readers of this book will find the opportunity to contribute to this research.

Symbols Used in Equations and Figures

Latent variables

- G additive genotype
- G′ additive genotype of parent when a child
- C common (shared) environment
- D genetic dominance
- I genetic interaction (epistasis)
- U unshared environment

Observed scores

- M mother
- F father
- A adopted child
- B biological child
- T twin

Paths

- h effect of additive genes on trait
- d effect of genetic dominance on trait
- i effect of epistasis on trait

u effect of unshared environment on trait
c effect of shared environment on trait
subscripts:
MZ monozygotic twin
DZ dizygotic twin
S sibling (sometimes includes DZs or all twins)
P parent and child
m or f added to subscript indicates sex: e.g., c^2_{MZf} = shared environmental variance of female MZs
m effect of mother's trait on shared environment of children
f effect of father's trait on shared environment of children
t genetic correlation over time (between G and G')

References

Abdel-Rahim, A. R., Nagoshi, C. T., Johnson, R. C., & Vandenberg, S. G. (1988). Familial resemblances for cognitive abilities and personality in an Egyptian sample. *Personality and Individual Differences, 9,* 155-163.

Ahern, F. M., Johnson, R. C., Wilson, J. R., McClearn, G. E., & Vandenberg, S. G. (1982). Family resemblances in personality. *Behavior Genetics, 12,* 261-280.

Allen, G. (1970). Within group and between group variation expected in human behavioral characters. *Behavior Genetics, 1,* 175-194.

Allport, G. W. (1937). *Personality: A psychological interpretation.* New York: Henry Holt.

Allport, G. W., & Odbert, H. S. (1936). Trait-names: A psycho-lexical study. *Psychological Monographs, 47* (Whole No. 211).

Bentler, P. M. (1989). *EQS structural equations program manual.* Los Angeles: BMDP Statistical Software.

Bergeman, C. S., & Chipuer, H. M. (1989). A twin/adoption study of the "little three" of the "big five" personality traits: Openness to experience, Agreeableness, and Conscientiousness. *Behavior Genetics, 19,* 744 (Abstract).

Bergeman, C. S., Chipuer, H. M., Plomin, R., Pedersen, N. L., McClearn, G. E., Nesselroade, J. R., Costa P., Jr., & McCrae, R. R. (in press). Genetic and environmental effects on openness to experience, agreeableness, and conscientiousness. An adoption/twin study. *Journal of Personality.*

Bergeman, C. S., Plomin, R., McClearn, G. E., Pedersen, N. L., & Friberg, L. T. (1988). Genotype-environment interaction in personality development: Identical twins reared apart. *Psychology and Aging, 3,* 399-406.

Boomsma, D. I., Martin, N. G., & Neale, M. C. (Eds.) (1989).Genetic analysis of twin and family data: Structural modeling using LISREL. *Behavior Genetics, 19*(1) (Special issue).

Bouchard, T. J., Jr., & McGue, M. (1990). Genetic and rearing environmental influences on adult personality: An analysis of adopted twins reared apart. *Journal of Personality, 58*, 263-292.

Boyle, G. J. (1989). Re-examination of the major personality-type factors in the Cattell, Comrey and Eysenck scales: Were the factors solutions by Noller et al. optimal? *Personality and Individual Differences, 10*, 1289-1299.

Braungart, J. M., Plomin, R, DeFries, J. C., & Fulker, D. (in press). Genetic influence on tester-rated infant temperament as assessed by Bayley's Infant Behavior Record: Nonadoptive and adoptive siblings and twins. *Developmental Psychology.*

Buss, A. H., & Plomin, R. (1975). *A temperament theory of personality development.* New York: Wiley.

Buss, D. M. (1984). Evolutionary biology and personality psychology. *American Psychologist, 39*, 1135-1147.

Buss, D. M. (1990). Toward a biologically informed psychology of personality. *Journal of Personality, 58*, 1-16.

Buss, D. M. (1991). Evolutionary personality psychology. *Annual Review of Psychology, 42*, 459-491.

Buss, D. M., & Craik, K. H. (1983). The act frequency approach to personality. *Psychological Review, 90*, 105-126.

Carey, G., & Rice, J. (1983). Genetics and personality temperament: Simplicity or complexity? *Behavior Genetics, 13*, 43-63.

Cattell, R. B. (1933). Temperament tests. I. Temperament. *British Journal of Psychology, 23*, 308-329.

Cattell, R. B. (1943). The description of personality: Basic traits resolved into clusters. *Journal of Abnormal and Social Psychology, 38*, 476-506.

Cattell, R. B. (1946). *Description and measurement of personality.* Yonkers-on-Hudson, NY: World.

Cattell, R. B. (1982). *The inheritance of personality and ability.* New York: Academic Press.

Cattell, R. B., Schuerger, J. M., & Klein, T. W. (1982). Heritabilities of Ego Strength (Factor C), Super Ego Strength (Factor G), and Self-Sentiment (Factor Q3) by multiple abstract variance analysis. *Journal of Clinical Psychology, 38*, 769-779.

Chen, C. J., Yu, M. W., Wang, C. J., Tong, S. L., & Tien, M. (1989). Genetic variance and heritability of temperament among Chinese twin infants. *Acta Geneticae Medicae et Gemellologiae, 38*, 215 (Abstract).

Costa, P. T., Jr., & McCrae, R. R. (1985). *The NEO Personality Inventory manual.* Odessa, FL: Psychological Assessment Resources.

Daly, M., & Wilson, M. (1990). Is parent-offspring conflict sex-linked? Freudian and Darwinian models. *Journal of Personality, 58*, 163-189.

Davidon, W. C. (1975). Optimally conditioned optimization algorithms without line searches. *Mathematical Programming, 9*, 1-30.

DeFries, J. C., & Fulker, D. W. (1985). Multiple regression analysis of twin data. *Behavior Genetics, 15*, 467-473.

Digman, J. M. (1990). Personality structure: Emergence of the five-factor model. *Annual Review of Psychology, 41*, 417-440.

Draper, P., & Belsky, J. (1990). Personality development in evolutionary perspective. *Journal of Personality, 58*, 141-161.

Dworkin, R. H., Burke, B. W., Maher, B. A., & Gottesman, I. I. (1976). A longitudinal study of the genetics of personality. *Journal of Personality and Social Psychology, 34*, 510-518.

Dworkin, R. H., Burke, B. W., Maher, B. A., & Gottesman, I. I. (1977). Genetic influences on the organization and development of personality. *Developmental Psychology, 13*, 164-165.

Eaves, L., & Eysenck, H. (1975). The nature of extraversion: A genetical analysis. *Journal of Personality and Social Psychology, 32*, 102-112.

Eaves, L., & Eysenck, H. (1976a). Genetic and environmental components of inconsistency and unrepeatability in twins' responses to a neuroticism questionnaire. *Behavior Genetics, 6*, 145-160.

Eaves, L., & Eysenck, H. (1976b). Genotype × age interaction for neuroticism. *Behavior Genetics, 6*, 359-362.

Eaves, L. J., Eysenck, H. J., & Martin, N. G. (1989). *Genes, culture and personality.* London: Academic Press.

Eaves, L. J., Martin, N. G., Heath, A. C., Hewitt, J. K., & Neale, M. C. (1990). Personality and reproductive fitness. *Behavior Genetics, 20*, 563-568.

Fiske, D. W. (1949). Consistency of the factorial structures of personality ratings from different sources. *Journal of Abnormal and Social Psychology, 44*, 329-344.

Floderus-Myrhed, B., Pedersen, N., & Rasmuson, I. (1980). Assessment of heritability for personality, based on a short-form of the Eysenck Personality Inventory. *Behavior Genetics, 10*, 153-162.

Fuller, J. L., & Thompson, W. R. (1978). *Foundations of behavior genetics.* St. Louis, MO: Mosby.

Gangestad, S. W., & Simpson, J. A. (1990). Toward an evolutionary history of female sociosexual variation. *Journal of Personality, 58*, 69-96.

Goldsmith, H. H., Buss, A. H., Plomin, R., Rothbart, M. K., Thomas, A., Chess, S., Hinde, R. A., & McCall, R. B. (1987). Roundtable: What is temperament? Four approaches. *Child Development, 58*, 505-529.

Goldsmith, H. H., & Campos, J. J. (1986). Fundamental issues in the study of early temperament: The Denver twin temperament study. In M. E. Lamb, A. L. Brown, & B. Rogoff (Eds.), *Advances in developmental psychology* (pp. 231-283). Hillsdale, NJ: Erlbaum.

Goldsmith, H. H., & Gottesman, I. I. (1981). Origins of variation in behavioral style: A longitudinal study of temperament in young twins. *Child Development, 52*, 91-103.

Gottesman, I. I. (1966). Genetic variance in adaptive personality traits. *Journal of Child Psychology and Psychiatry, 7*, 199-208.

Hay, D. A. (1985). *Essentials of behaviour genetics.* Melbourne: Blackwell.

Heath, A. C., Jardine, R., Eaves, L. J., & Martin, N. G. (1989). The genetic structure of personality. II. Genetic item analysis of the EPQ. *Personality and Individual Differences, 10*, 615-624.

Hewitt, J. K. (1984). Normal components of personality variation. *Journal of Personality and Social Psychology, 47*, 671-675.

Horn, J. M., Loehlin, J. C., & Willerman, L. (1979). Intellectual resemblance among adoptive and biological relatives: The Texas Adoption Project. *Behavior Genetics, 9*, 177-207.

Horn, J. M., Plomin, R., & Rosenman, R. (1976). Heritability of personality traits in adult male twins. *Behavior Genetics, 6*, 17-30.

Jinks, J. L., & Fulker, D. W. (1970). Comparison of the biometrical genetical, MAVA, and classical approaches to the analysis of human behavior. *Psychological Bulletin, 73,* 311-349.

John, O. P. (1990). The "Big Five" factor taxonomy: Dimensions of personality in the natural languages and in questionnaires. In L. A. Pervin (Ed.), *Handbook of personality: Theory and research* (pp. 66-100). New York: Guilford Press.

John, O. P., Goldberg, L. R., & Angleitner, A. (1984). Better than the alphabet: Taxonomies of personality-descriptive terms in English, Dutch, and German. In H. Bonarius, G. van Heck, & N. Smid (Eds.), *Personality psychology in Europe: Theoretical and empirical developments* (pp. 83-100). Lisse, The Netherlands: Swets & Zeitlinger.

Jöreskog, K. G., & Sörbom, D. (1989). *LISREL 7: A guide to the program and applications* (2nd ed.). Chicago: SPSS.

Jung, C. G. (1924). *Psychological types.* New York: Harcourt Brace.

Kaprio, J., Koskenvuo, M., & Rose, R. J. (1990). Change in cohabitation and intrapair similarity of monozygotic (MZ) twins for alcohol use, extraversion, and neuroticism. *Behavior Genetics, 20,* 265-276.

Lange, K. L., Westlake, J., & Spence, M. A. (1976). Extensions to pedigree analysis. III. Variance components by the scoring method. *Annals of Human Genetics, 39,* 485-491.

Langinvainio, H., Kaprio, J., Koskenvuo, M., & Lönnqvist, J. (1984). Finnish twins reared apart. III: Personality factors. *Acta Geneticae Medicae et Gemellologiae, 33,* 259-264.

Little, B. R. (1989). Personal projects analysis. In D. M. Buss & N. Cantor (Eds.), *Personality psychology: Recent trends and emerging directions* (pp. 15-31). New York: Springer-Verlag.

Loehlin, J. C. (1965). A heredity-environment analysis of personality inventory data. In S. G. Vandenberg (Ed.), *Methods and goals in human behavior genetics* (pp. 163-168). New York: Academic Press.

Loehlin, J. C. (1986a). Are California Psychological Inventory *items* differently heritable? *Behavior Genetics, 16,* 599-603.

Loehlin, J. C. (1986b). Heredity, environment, and the Thurstone Temperament Schedule. *Behavior Genetics, 16,* 61-73.

Loehlin, J. C. (1987). *Latent variable models: An introduction to factor, path, and structural analysis.* Hillsdale, NJ: Erlbaum.

Loehlin, J. C., & DeFries, J. C. (1987). Genotype-environment correlation and IQ. *Behavior Genetics, 17,* 263-277.

Loehlin, J. C., & Gough, H. G. (1990). Genetic and environmental variation on the California Psychological Inventory vector scales. *Journal of Personality Assessment, 54,* 463-468.

Loehlin, J. C., Horn, J. M., & Willerman, L. (1981). Personality resemblance in adoptive families. *Behavior Genetics, 11,* 309-330.

Loehlin, J. C., Horn, J. M., & Willerman, L. (1989). Modeling IQ change: Evidence from the Texas Adoption Project. *Child Development, 60,* 993-1004.

Loehlin, J. C., Horn, J. M., & Willerman, L. (1990). Heredity, environment, and personality change: Evidence from the Texas Adoption Project. *Journal of Personality, 58,* 221-243.

Loehlin, J. C., & Nichols, R. C. (1976). *Heredity, environment, and personality.* Austin, TX: University of Texas Press.

Loehlin, J. C., Willerman, L., & Horn, J. M. (1982). Personality resemblances between unwed mothers and their adopted-away offspring. *Journal of Personality and Social Psychology, 42*, 1089-1099.

Loehlin, J. C., Willerman, L., & Horn, J. M. (1985). Personality resemblances in adoptive families when the children are late-adolescent or adult. *Journal of Personality and Social Psychology, 48*, 376-392.

Loehlin, J. C., Willerman, L., & Horn, J. M. (1987). Personality resemblance in adoptive families: A 10-year follow-up. *Journal of Personality and Social Psychology, 53*, 961-969.

Lykken, D. T. (1982). Research with twins: The concept of emergenesis. *Psychophysiology, 19*, 361-373.

Lykken, D. T., McGue, M., Bouchard, T. J., Jr., & Tellegen, A. (1990). Does contact lead to similarity or similarity to contact? *Behavior Genetics, 20*, 547-561.

Lykken, D. T., Tellegen, A., & DeRubeis, R. (1978). Volunteer bias in twin research: The rule of two-thirds. *Social Biology, 25*, 1-9.

Martin, N. & Jardine, R. (1986). Eysenck's contributions to behaviour genetics. In S. Modgil & C. Modgil (Eds.), *Hans Eysenck: Consensus and controversy* (pp. 13-47). Philadelphia: Falmer Press.

Matheny, A. P., Jr. (1980). Bayley's Infant Behavior Record: Behavioral components and twin analyses. *Child Development, 51*, 1157-1167.

Matheny, A. P., Jr. (1983). A longitudinal twin study of stability of components from Bayley's Infant Behavior Record. *Child Development, 53*, 356-360.

Matheny, A. P., Jr. (1987). Developmental research of twin's temperament. *Acta Geneticae Medicae et Gemellologiae, 36*, 135-143.

Matheny, A. P., Jr., Wilson, R. S., & Dolan, A. B. (1976). Relations between twins' similarity of appearance and behavioral similarity. *Behavior Genetics, 6*, 343-351.

McArdle, J. J. (1986). Latent variable growth within behavior genetic models. *Behavior Genetics, 16*, 163-200.

McCartney, K., Harris, M. J., & Bernieri, F. (1990). Growing up and growing apart: A developmental meta-analysis of twin studies. *Psychological Bulletin, 107*, 226-237.

McCrae, R. R. (1989). Why I advocate the five-factor model: Joint factor analyses of the NEO-PI with other instruments. In D.M. Buss & N. Cantor (Eds.), *Personality psychology: Recent trends and emerging directions* (pp. 237-245). New York: Springer-Verlag.

McCrae, R. R., & Costa, P. T., Jr. (1985). Updating Norman's "adequate taxonomy": Intelligence and personality dimensions in natural language and in questionnaires. *Journal of Personality and Social Psychology, 49*, 710-721.

McCrae, R. R., & Costa, P. T., Jr. (1987). Validation of the five-factor model of personality across instruments and observers. *Journal of Personality and Social Psychology, 52*, 81-90.

McGue, M., Wette, R., & Rao, D. C. (1984). Evaluation of path analysis through computer simulation: Effect of incorrectly assuming independent distribution of familial correlations. *Genetic Epidemiology, 1*, 255-269.

Mealey, L. (1990). Differential use of reproductive strategies by human groups? *Psychological Science, 1*, 385-387.

Murray, H. A. (1938). *Explorations in personality.* New York: Oxford University Press.

Neale, M. C., Rushton, J. P., & Fulker, D. W. (1986). Heritability of item responses on the Eysenck Personality Questionnaire. *Personality and Individual Differences, 7*, 771-779.

Norman, W. T. (1963). Toward an adequate taxonomy of personality attributes: Replicated factor structure in peer nomination personality ratings. *Journal of Abnormal and Social Psychology, 66,* 574-583.

Osborne, R. T. (1980). *Twins: Black and white.* Athens, GA: Foundation for Human Understanding.

Pedersen, N. L., Gatz, M., Plomin, R., Nesselroade, J. R., & McClearn, G. E. (1989). Individual differences in locus of control during the second half of the life span for identical and fraternal twins reared apart and reared together. *Journal of Gerontology: Psychological Sciences, 44,* 100-105.

Pedersen, N. L., Lichtenstein, P., Plomin, R., DeFaire, U., McClearn, G. E., & Matthews, K. A. (1989). Genetic and environmental influences for Type A-like measures and related traits: A study of twins reared apart and twins reared together. *Psychosomatic Medicine, 51,* 428-440.

Pedersen, N. L., Plomin, R., McClearn, G. E., & Friberg, L. (1988). Neuroticism, Extraversion, and related traits in adult twins reared apart and reared together. *Journal of Personality and Social Psychology, 55,* 950-957.

Plomin, R. (1986). *Development, genetics, and psychology.* Hillsdale, NJ: Erlbaum.

Plomin, R., Coon, H., Carey, G., DeFries, J. C., & Fulker, D. W. (1991). Parent-offspring and sibling adoption analyses of parental ratings of temperament in infancy and childhood. *Journal of Personality, 59,* 705-732.

Plomin, R., & Daniels, D. (1987). Why are children in the same family so different from one another? *Behavioral and Brain Sciences, 10,* 1-16.

Plomin, R., & DeFries, J. C. (1985). *Origins of individual differences in infancy: The Colorado Adoption Project.* Orlando, FL: Academic Press.

Plomin, R., DeFries, J. C., & Fulker, D. W. (1988). *Nature and nurture during infancy and early childhood.* Cambridge, UK: Cambridge University Press.

Plomin, R., DeFries, J. C., & Loehlin, J. C. (1977). Genotype-environment interaction and correlation in the analysis of human behavior. *Psychological Bulletin, 84,* 309-322.

Plomin, R., DeFries, J. C., & McClearn, G. E. (1990). *Behavioral genetics: A primer* (2nd ed.). New York: W. H. Freeman.

Plomin, R., & McClearn, G. E. (1990). Human behavioral genetics of aging. In J. E. Birren & K. W. Schaie (Eds.), *Handbook of the psychology of aging* (3rd ed., pp. 67-78). San Diego: Academic Press.

Plomin, R., McClearn, G. E., Pedersen, N. L., Nesselroade, J. R., & Bergeman, C. S. (1988). Genetic influence on childhood family environment perceived retrospectively from the last half of the life span. *Developmental Psychology, 24,* 738-745.

Plomin, R., Pedersen, N. L., McClearn, G. E., Nesselroade, J. R., & Bergeman, C. S. (1988). EAS temperaments during the last half of the life span: Twins reared apart and twins reared together. *Psychology and Aging, 3,* 43-50.

Plomin, R., Willerman, L., & Loehlin, J. C. (1976). Resemblance in appearence and the equal environments assumption in twin studies of personality traits. *Behavior Genetics, 6,* 43-52.

Pogue-Geile, M. F., & Rose, R. J. (1985). Developmental genetic studies of adult personality. *Developmental Psychology, 21,* 547-557.

Price, R. A., Vandenberg, S. G., Iyer, H., & Williams, J. S. (1982). Components of variation in normal personality. *Journal of Personality and Social Psychology, 43,* 328-340.

Rahe, R. H., Hervig, L., & Rosenman, R. H. (1978). Heritability of Type A behavior. *Psychosomatic Medicine, 40,* 478-486.

Rao, D. C., Morton, N. E., Elston, R. C., & Yee, S. (1977). Causal analysis of academic performance. *Behavior Genetics, 7,* 147-159.

Riese, M. L. (1990). Neonatal temperament in monozygotic and dizygotic twin pairs. *Child Development, 61,* 1230-1237.

Rose, R. J. (1988). Genetic and environmental variance in content dimensions of the MMPI. *Journal of Personality and Social Psychology, 55,* 302-311.

Rose, R. J., & Kaprio, J. (1988). Frequency of social contact and intrapair resemblance of adult monozygotic cotwins. *Behavior Genetics, 18,* 309-328.

Rose, R. J., & Kaprio, J., with Williams, C. J., Viken, R., & Obremski, K. (1990). Social contact and sibling similarity: Facts, issues, and red herrings. *Behavior Genetics, 20,* 763-778.

Rose, R. J., Koskenvuo, M., Kaprio, J., Sarna, S., & Langinvainio, H. (1988). Shared genes, shared experiences, and similarity of personality. *Journal of Personality and Social Psychology, 54,* 161-171.

Rosenman, R. H., Rahe, R. H., Borhani, N. O., & Feinleib, M. (1976). Heritability of personality and behavior pattern. *Acta Geneticae Medicae et Gemellologiae, 25,* 221-224.

Rowe, D. C. (1981). Environmental and genetic influences on dimensions of perceived parenting: A twin study. *Developmental Psychology, 17,* 203-208.

Rowe, D. C. (1983). A biometrical analysis of perceptions of family environments: A study of twin and singleton sibling kinships. *Child Development, 54,* 416-423.

Rowe, D. C., Clapp, M., & Wallis, J. (1987). Physical attractiveness and the personality resemblance of identical twins. *Behavior Genetics, 17,* 191-201.

Rushton, J. P. (1985). Differential K theory: The sociobiology of individual and group differences. *Personality and Individual Differences, 6,* 441-452.

Rushton, J. P., Fulker, D. W., Neale, M. C., Nias, D. K. B., & Eysenck, H. J. (1986). Altruism and aggression: The heritability of individual differences. *Journal of Personality and Social Psychology, 50,* 1192-1198.

Sarbin, T. R., & Allen, V. L. (1968). Role theory. In G. Lindzey & E. Aronson (Eds.), *The handbook of social psychology* (2nd ed., vol. 1., pp. 488-567). Reading, MA: Addison-Wesley.

Scarr, S., & McCartney, K. (1983). How people make their own environments: A theory of genotype → environment effects. *Child Development, 54,* 424-435.

Scarr, S., Webber, P. L., Weinberg, R. A., & Wittig, M. A. (1981). Personality resemblance among adolescents and their parents in biologically related and adoptive families. *Journal of Personality and Social Psychology, 40,* 885-898.

Shields, J. (1962). *Monozygotic twins: Brought up apart and brought up together.* London: Oxford University Press.

Tellegen, A., Lykken, D. T., Bouchard, T. J., Jr., Wilcox, K. J., Segal, N. L., & Rich, S. (1988). Personality similarity in twins reared apart and together. *Journal of Personality and Social Psychology, 54,* 1031-1039.

Thiessen, D. D. (1972). A move toward species-specific analysis in behavior genetics. *Behavior Genetics, 2,* 115-126.

Tooby, J., & Cosmides, L. (1990). On the universality of human nature and the uniqueness of the individual: The role of genetics and adaptation. *Journal of Personality, 58,* 17-67.

Vandenberg, S. G. (1962). The hereditary abilities study: Hereditary components in a psychological test battery. *American Journal of Human Genetics, 14,* 220-237.

Wilson, E. O. (1975). *Sociobiology: The new synthesis.* Cambridge, MA: Belknap Press.

Wilson, E. O. (1978). *On human nature.* Cambridge, MA: Harvard University Press.

Wilson, R. S. (1978). Synchronies in mental development. *Science, 202,* 939-948.

Wilson, R. S., & Matheny, A. P., Jr. (1986). Behavior-genetics research in infant temperament: The Louisville Twin Study. In R. Plomin & J. Dunn (Eds.), *The Study of Temperament* (pp. 81-97). Hillsdale, NJ: Erlbaum.

Index

About the Author

John C. Loehlin received his Ph.D. in Psychology in 1957 from the University of California at Berkeley. He taught at the University of Nebraska until 1964; since then he has taught at The University of Texas at Austin except for a year as a Fellow at the Center for Advanced Study in the Behavioral Sciences.